BLUEPRINT READING FOR MACHINISTS
INTERMEDIATE
Fourth Edition

BLUEPRINT READING FOR MACHINISTS
INTERMEDIATE
Fourth Edition
David L. Taylor

DELMAR PUBLISHERS INC.

NOTICE TO THE READER

Publisher does not warrant or guarantee any of the products described herein or perform any independent analysis in connection with any of the product information contained herein. Publisher does not assume, and expressly disclaims, any obligation to obtain and include information other than that provided to it by the manufacturer.

The reader is expressly warned to consider and adopt all safety precautions that might be indicated by the activities described herein and to avoid all potential hazards. By following the instructions contained herein, the reader willingly assumes all risks in connection with such instructions.

The publisher makes no representations or warranties of any kind, including but not limited to, the warranties of fitness for particular purpose or merchantability, nor are any such representations implied with respect to the material set forth herein, and the publisher takes no responsibility with respect to such material. The publisher shall not be liable for any special, consequential or exemplary damages resulting, in whole or in part, from the readers' use of, or reliance upon, this material.

Cover photo courtesy of THE FERGUSON DRIVE®
A product of the Ferguson Machine Company,
Division of MNC Industries, Inc.
for high speed precision intermittent motion.

For information, address Delmar Publishers Inc.
2 Computer Drive West, Box 15-015
Albany, New York 12212

Copyright © 1984 by Delmar Publishers Inc.

All rights reserved. Certain portions of this work copyright 1971. No part of this work may be reproduced or used in any form or by any means — graphic, electronic, or mechanical, including photocopying, recording, taping, or information storage and retrieval systems — without written permission of the publisher.

Printed in the United States of America
Published simultaneously in Canada
by Nelson Canada.
a division of International Thomson Limited

10 9 8 7 6 5

Library of Congress Catalog Card Number 84-12670
ISBN 0-8273-1085-4

CONTENTS

Preface ix

Unit 1   ORTHOGRAPHIC PROJECTION 1
- Principles of Projection 1
- Arrangement of Views 4
- Application of Projection Lines 5
- Angles of Orthographic Projection 5
- Specifying Repetitive Features 8
- Preparation of Drawings Using Computer-
  - aided Design (CAD) Systems 8
  - Assignment D-1 – SEPARATOR BRACKET 11
  - Assignment D-2 – BASE PLATE 13

Unit 2   SECTION VIEWS 14
- Cutting Plane Lines 14
- Offset Cutting Planes 17
- Section Lines 17
- Full Sections 17
- Half Sections 18
- Revolved Sections 18
- Removed Sections 18
- Broken-out Sections 18
  - Assignment D-3 – RAISE BLOCK 21
  - Assignment D-4 – SLIDE VALVE 23

Unit 3   SURFACE TEXTURE 24
- Surface Texture Terminology 24
- Surface Texture Symbols 25
- Lay Symbols 27
- Measuring Surface Texture 27
  - Assignment D-5 – SHAFT INTERMEDIATE SUPPORT 31

Unit 4   VIOLATIONS OF TRUE PROJECTION 32
- Rotated or Aligned Projection 32
- Untrue Projection 33
- Untrue Projection of Sections 34
  - Assignment D-6 – SPARK ADJUSTER 37
  - Assignment D-7 – COIL FRAME 39

Unit 5	SPECIAL VIEWS	41
	Partial Views	41
	Distorted Views	42
	Bottom Views	43
	Phantom Lines and Views	44
	Assignment D-8 – INDEX PEDESTAL	47
	Assignment D-9 – YOKE	49
Unit 6	POSITIONAL DIMENSIONING	51
	Point-to-point Dimensions	51
	Datum Dimensioning	51
	Assignment D-10 – INTERLOCK BASE	55
	Assignment D-11 – CASE COVER	57
Unit 7	GEOMETRIC TOLERANCES – DATUMS	58
	Terminology	58
	Datums	59
	Datum Plane	59
	Datum Cylinder	61
	Datum Axis	61
	Datum Identification Symbol	61
	Feature Control Symbols	61
	Assignment D-12 – POSITIONING ARM	65
Unit 8	GEOMETRIC TOLERANCES – LOCATION AND FORM	67
	Modifiers	67
	Location Tolerances	68
	Form Tolerances	69
	Review of Symbology	71
	Assignment D-13 – TRIP BOX	75
Unit 9	SCREW THREADS	77
	Screw Thread Forms	77
	Screw Thread Terminology	78
	Unified National Thread Series	80
	Multiple Threads	81
	Classification of Fits	81
	Symbols for Identifying Thread Specifications	82
	Pictorial Representation	82
	Schematic Representation	82
	Simplified Thread Representation	83
	Representing Tapped Holes	83
	Assignment D-14 – CROSS HEAD	85
	Assignment D-15 – SPINDLE BEARING	87

Unit 10	THREADED FASTENERS		88
	Machine Screws		88
	Cap Screws		88
	Machine Bolts		89
	Stud Bolts		89
	Setscrews		90
	Washers		90
	Threaded Fastener Size		90
		Assignment D-16 – SPIDER	95
		Assignment D-17 – FLANGE	97
Unit 11	PIPE THREADS		99
	American National Standard		99
	Representation of Pipe Threads		100
		Assignment D-18 – DRIVE HOUSING	103
Unit 12	IDENTIFYING STEELS		104
	AISI and SAE Systems		104
		Assignment D-19 – REAR TOOL POST	109
Unit 13	DOVETAILS		111
	Description of Dovetails		111
	Measuring Dovetails		112
		Assignment D-20 – SHUTTLE	115
		Assignment D-21 – DRILL SLIDE	117
Unit 14	CASTING		118
	Sand Molding		118
	Flat Back Patterns		119
	Coring		121
	Cored Castings		122
		Assignment D-22 – AUXILIARY PUMP BASE	125
Unit 15	FINISHES AND PROTECTIVE COATINGS		126
	Conversion Coating		126
	Electroplating		127
	Flame Spray Coating		127
	Organic Coatings		127
		Assignment D-23 – CORNER BRACKET	129
Unit 16	STRUCTURAL STEEL SHAPES		131
	Common Shapes		131
	Shape Designations		131
	Built-up Sections		133
		Assignment D-24 – FOUR WHEEL TROLLEY	135

Unit 17	WELDING	137
	Welding Joints	137
	Types of Welds	137
	Weld and Welding Symbols	139
	Terminology	139
	Location of Welding Symbols	141
	Assignment D-25 – SUPPORT ASSEMBLY VALVE	143
	Assignment D-26 – STOCK PUSHER GUIDE	145
Unit 18	PIN FASTENERS	146
	Taper Pins	146
	Dowel Pins	148
	Straight Pins	148
	Grooved Pins	150
	Clevis Pins	151
	Spring Pins	151
	Cotter Pins	155
	Assignment D-27 – HOOD	157
Unit 19	SPRINGS	158
	Helical Springs	159
	Flat Springs	160
	Specifications of Springs on Drawings	161
	Assignment D-28 – FLUID PRESSURE VALVE	163
Unit 20	SWIVELS, UNIVERSAL JOINTS, AND BEARINGS	164
	Swivels	164
	Universal Joint	164
	Boring Split Bearings	165
	Assignment D-29 – UNIVERSAL TROLLEY	167
Unit 21	WORM GEARING	168
	Representation of Worm Gears	168
	Worm Gearing Parts, Symbols, and Terms	169
	Assignment D-30 – WORM GEAR	173
	Assignment D-31 – WORM SPINDLE	175

Appendix	176
Illustration Contributions	182
Index	183

PREFACE

To meet the challenge of modern manufacturing requirements, students must perfect their skills in reading and interpreting a variety of working drawings. Metalworking, quality control, product engineering, process planning for numerical control, computer programming for computer-aided drafting and manufacturing systems, and inspection are just some of the careers which involve extensive use of technical drawings. Students who possess the needed skills will find rewarding job opportunities.

The fourth edition of *Blueprint Reading for Machinists–Intermediate* is designed to follow a basic blueprint reading course in which students learn the rudiments of interpreting machine drawings. Given this background, the student can move into the graded instruction presented in the intermediate text.

Blueprint Reading for Machinists–Intermediate helps students develop the required skills by presenting information in a logical progression of units, moving from easy to more difficult principles. Immediate reinforcement of the understanding obtained from studying each unit is provided by the assignment drawing and the related questions. The information provided in the unit will enable the student to answer all of the questions; additional references are not required. As students master new principles and perfect their interpretive skills, the drawings keep pace by providing increasingly more challenging assignments.

Content retained from previous editions of the text has been reorganized, revised and upgraded to reflect the requirements of the latest standards, including ANSI Y14.5M-1982. New topics have been added to provide a broader coverage of principles and to ensure that students are well-prepared for the job market. Some of the new or expanded topics included are:

- Datums for geometric tolerancing
- Geometric tolerances of location and form
- Symbology for geometric dimensioning and tolerancing
- Threaded fasteners
- Finishes and protective coatings
- Structural steel shapes
- Welding symbols, joints and terminology (reflecting the latest American Welding Society standard)
- Pin fasteners

The number of assignment drawings has been increased to provide additional opportunities for students to practice and perfect their interpretive skills.

The Appendix section of the text gives useful handbook tables for reference. An Instructor's Guide is available to accompany the text. The Guide contains answers to each assignment question given in the text.

ABOUT THE AUTHOR

David L. Taylor is a team leader for the Cummins Engine Company and journeyman tool and die maker. He holds a degree in Vocational-technical Education from the State University of New York at Buffalo and has taught blueprint reading to secondary school vocational students and to students in community college programs. He is the author of several Delmar texts, including *Drill Press Work, Elementary Blueprint Reading for Machinists,* and *Machine Trades Blueprint Reading.*

unit 1

ORTHOGRAPHIC PROJECTION

Industrial drawings and prints furnish a description of the shape and size of an object. All information necessary for its construction must be presented in a form that is easily recognized. For this reason, a number of views are necessary. Each view shows a part of the object as it is seen by looking directly at each one of the surfaces. The method used to project these views on paper is called *orthographic projection*. When all the notes, symbols, and dimensions are added to the projected views, it becomes a working drawing. A working drawing supplies all the information required to construct the part, Figure 1-1.

The ability to interpret a drawing accurately is based on the mastery of two skills. The print reader must:

1. Know and understand certain standardized signs and symbols.
2. Visualize the completed object by examining the drawing itself.

Visualizing is the process of forming a mental picture of an object. It is the secret of successful drawing interpretation. Visualization requires an understanding of the exact relationship of the views to each other. It also requires a working knowledge of how the individual views are obtained through projection. When these views are connected mentally, the object has length, width, and thickness.

PRINCIPLES OF PROJECTION

Most objects can be drawn by projecting them onto the sheet in some combination of the front, top, and right-side views. To project the views of an object into the three views, imagine it placed in a square box with transparent sides, Figure 1-2. The top is hinged to swing directly over the front. The right side is hinged to swing directly to the right of the front.

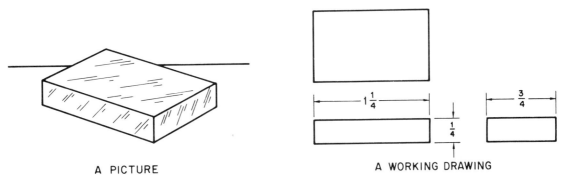

A PICTURE   A WORKING DRAWING

Fig. 1-1 Projecting an object into three views

2 Unit 1 Orthographic Projection

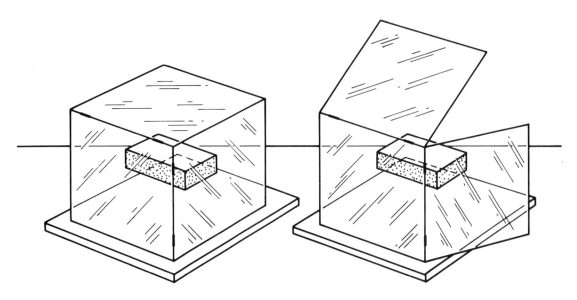

Fig. 1-2 Box with transparent sides

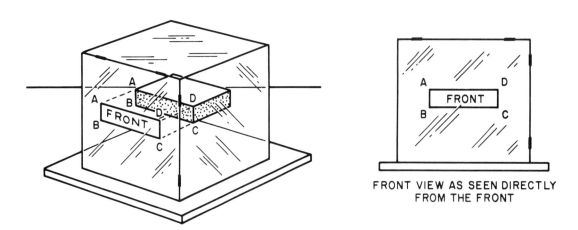

FRONT VIEW AS SEEN DIRECTLY FROM THE FRONT

Fig. 1-3 Projecting the front view

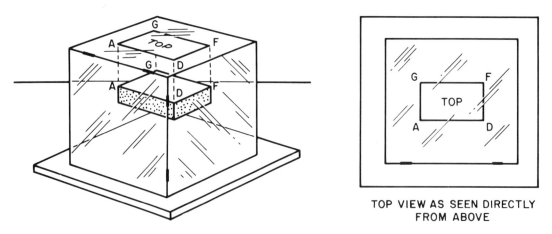

TOP VIEW AS SEEN DIRECTLY FROM ABOVE

Fig. 1-4 Projecting the top view

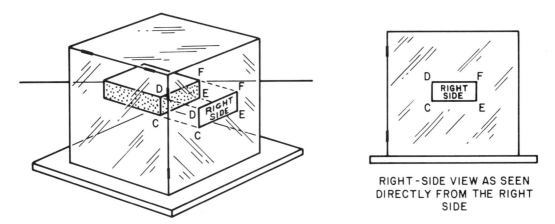

RIGHT-SIDE VIEW AS SEEN DIRECTLY FROM THE RIGHT SIDE

Fig. 1-5 Projecting the right-side view

In this case, the surfaces of the object selected are rectangular in shape. The front surface of the object is placed parallel to the front surface of the box. With the object held in this position, the outline of its front surface is traced on the face of the box as it would appear to the observer looking directly at it.

Note that the front of the object, indicated by A, B, C, D as drawn on the front surface in its correct shape, shows only the length and thickness, Figure 1-3.

Without moving the object, the operation is repeated with the observer looking directly down at the top. Note that the top of the object, indicated by A, D, F, G as drawn on the top surface in its correct shape, shows only the length and width, Figure 1-4.

The operation is repeated again for the side view with the observer looking directly at the right side. Note that the right side of the object, indicated by D, C, E, F as drawn on the right side surface in its correct shape, shows only the width and thickness, Figure 1-5.

If the top of the box is swung upward to a vertical position, the top view would appear directly over the front view. If the right side of the box is swung forward, the side view would appear to the right and in line with the front view, Figure 1-6.

The sides of the box and the identification letters are now removed, leaving the three projected views of the object in their correct relation, Figure 1-7.

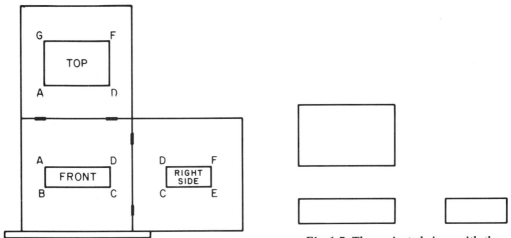

Fig. 1-6 The correct relation of the three views

Fig. 1-7 The projected views with the projection aids removed

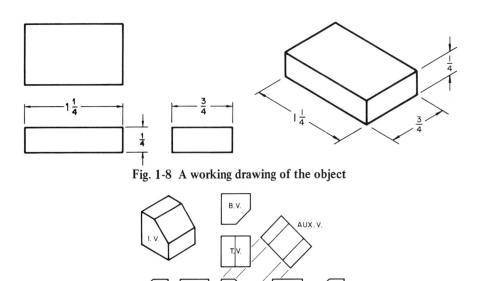

Fig. 1-8 A working drawing of the object

Fig. 1-9 Relative positions of views

A drawing has now been made of each of the three principal views (the front, the top, and the right side). Each shows the exact shape and size of the object and the relationship of the three views to each other. This principle of projection is used throughout all mechanical drawing.

To complete the drawing, dimensions and other information are added to the projected views. The drawing then becomes a *working drawing*. Such drawings furnish the information necessary for the construction of an object, Figure 1-8.

ARRANGEMENT OF VIEWS

The three-view drawing illustrated in Figure 1-7 shows the relative positions of the top, front, and right-side views. Often more or fewer views are needed to explain all the details of the part. The shape and the complexity of the object determines the number and arrangement of views. The drafter should supply enough detailed views of information for the construction of the object. Part of a drafter's job is to decide which views will best accomplish this purpose.

Figure 1-9 shows the position of views as they might appear on a working drawing. The name and location of each view is identified throughout this text as follows:

Front view	F.V.
Top view	T.V.
Right-side View	R.V.
Left-side View	L.V.
Bottom View	Bot. V.
Auxiliary View	Aux. V.
Back or Rear View	B.V.

The back view may be located in any one of the places indicated in Figure 1-9.

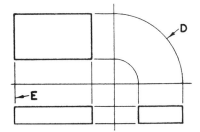 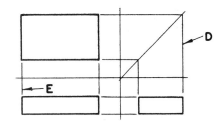

Fig. 1-10 Use of projection lines

APPLICATION OF PROJECTION LINES

The object in Figure 1-10 is shown with projection lines applied. *Projection lines* are used to show the relationship of points and surfaces in one view to those in other views. Ordinarily, the projection lines do not appear on a finished drawing. However, they may be found on drawings where a part is complicated. In this case, the projection lines show how certain details in a view were obtained. Practice in projecting lines in their imaginary positions from view to view help to develop skill in interpreting drawings.

Figure 1-11 shows a more complex object with projection lines applied. Note in this case that the corresponding points in each view have been lettered.

ANGLES OF ORTHOGRAPHIC PROJECTION

Third Angle Projection

Third angle projection is the recognized standard in the United States. This system places the projection or viewing plane between the object and the observer, Figure 1-12.

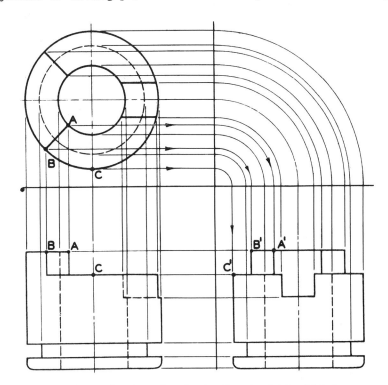

Fig. 1-11 Application of projection lines to obtain right view

6 Unit 1 Orthographic Projection

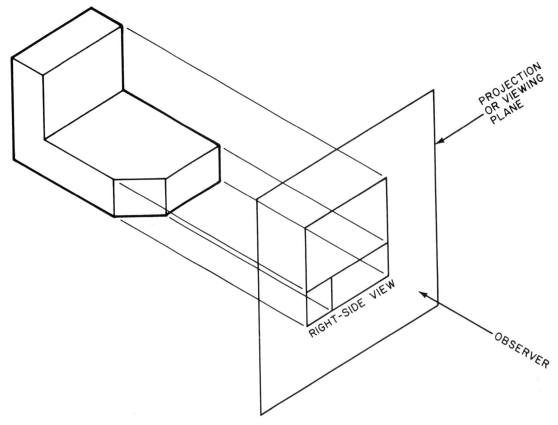

Fig. 1-12 In third angle projection, the projection (viewing) plane is between the object and the observer

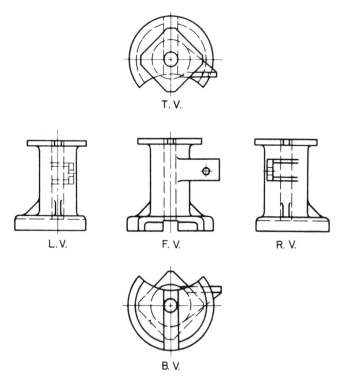

Fig. 1-13 Relative position of views in third angle projection (from C. Jensen & R. Hines, *Interpreting Engineering Drawings.* © 1970 by Delmar Publishers Inc.)

Unit 1 Orthographic Projection 7

The relative position of views in third angle projection is shown in Figure 1-13. Third angle projection is used in this text.

First Angle Projection

First angle projection is used in many countries of the world. In first angle projection the object is placed between the observer and the projection or viewing plane. Figure 1-14 shows the relative position of views in first angle projection.

The views in first angle projection are identical to those used in third angle projection. The difference between the two systems is the relative position of views in relation to the front view.

ISO Projection Symbols

To indicate the type of projection used on a drawing, a symbol is used. This symbol, which was developed by the International Standards Organization (ISO) is found in the title block area of the drawing. Figure 1-15 shows the standard ISO symbol used on drawings.

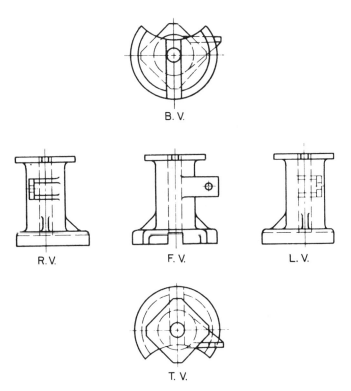

Fig. 1-14 First angle projection (from C. Jensen & R. Hines, *Interpreting Engineering Drawings.* © 1970 by Delmar Publishers Inc.)

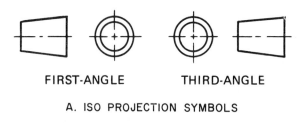

A. ISO PROJECTION SYMBOLS

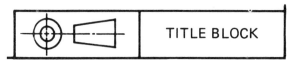

B. LOCATING ISO SYMBOL ON DRAWING PAPER

Fig. 1-15 Standard ISO projection symbols (from C. Jensen & R. Hines, *Interpreting Engineering Drawings.* © 1970 by Delmar Publishers Inc.)

SPECIFYING REPETITIVE FEATURES

Repetitive features or dimensions are often specified in more than one place on a drawing. To eliminate the need of dimensioning each individual feature, notes or symbols may be added to show that a process or dimension is repeated.

Holes of equal size may be called out by specifying the number of features required by an X. A small space is left between the X and the feature size dimension which follows, Figure 1-16.

Equal spacing of features such as hole patterns or slots may also be specified in a similar manner. The required number of typical features is called out followed by an X. The spacing dimension, usually expressed in degrees, is indicated following the X, Figure 1-17.

This new method, developed by the American National Standards Institute (ANSI), replaces the method previously used as shown in Figure 1-18.

PREPARATION OF DRAWINGS USING COMPUTER-AIDED DESIGN (CAD) SYSTEMS

The most common industrial drawings are produced by a drafter or designer using drafting instruments. However, the use of modern computer technology has greatly reduced drawing time by aiding the drafter in drawing production. Computer-aided design (CAD) systems are capable of automating many repetitive, time-consuming drawing tasks. An example of a computer-generated manufacturing drawing is provided in the assignment drawing for the Base Plate. The latest technology enables the drafter to reproduce drawings upon command to any given size or view. Three-dimensional qualities may also be given to the part and may be used in technical illustration.

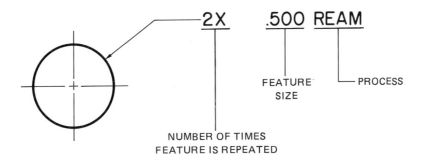

Fig. 1-16

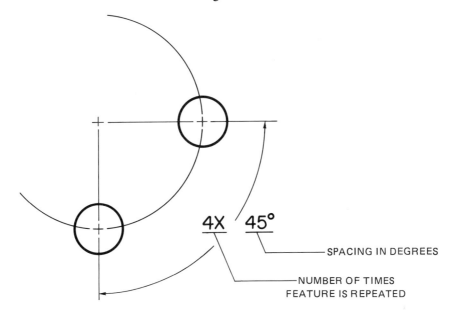

Fig. 1-17

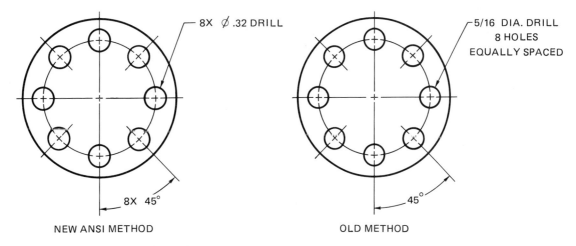

Fig. 1-18

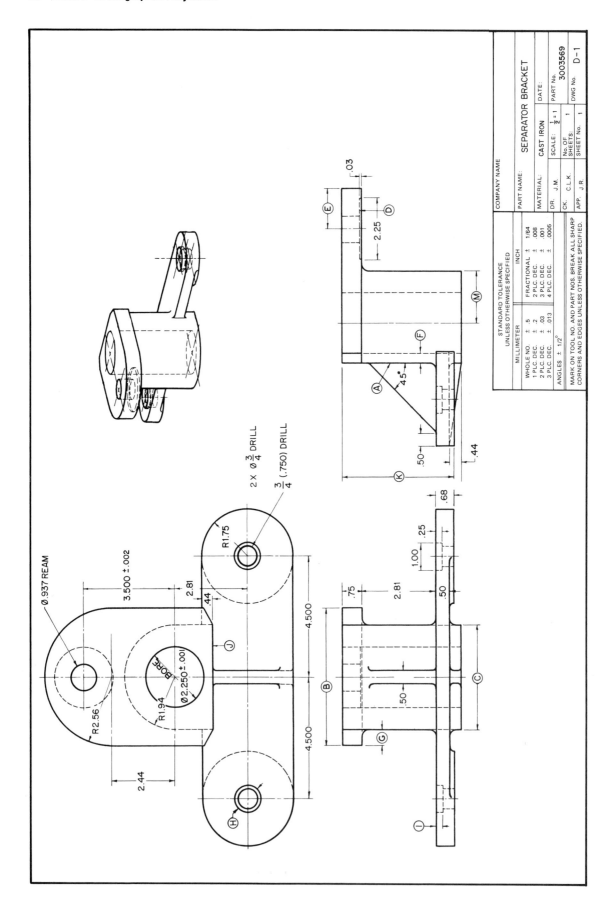

ASSIGNMENT D-2: BASE PLATE

1. How many 11/16 dia slots are required? _____
2. Determine distance Ⓐ . _____
3. How thick is the base plate? _____
4. Determine dimension Ⓑ . _____
5. Determine radius Ⓒ . _____
6. What is the upper limit dimension for the .500 slot? _____
7. What is the depth of the .75 slot? _____
8. What is the angle to which the two .500 slots are placed? _____
9. Determine distance Ⓓ . _____
10. Determine distance Ⓔ . _____

unit 2

SECTION VIEWS

The details of the interior of an object may be shown more clearly if a section view is provided. A *section view* is drawn as though a part of the object were cut away. This exposes the inside surfaces and makes the drawing easier to understand.

CUTTING PLANE LINES

The cutting plane line is used to indicate where a section of an object is taken, Figures 2-1 and 2-2. Figure 2-2 shows how to locate a cutting plane line. At C the arrows on the cutting plane line indicate the direction in which to look to obtain the sectioned view. At D, the letters A-A are used to help identify the sectioned view. On simple symmetrical objects, the cutting plane line may be omitted as at E.

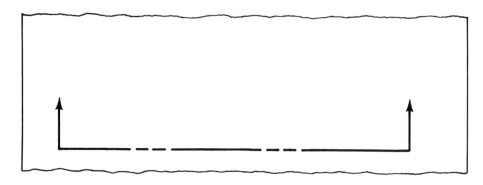

Fig. 2-1 Cutting plane line

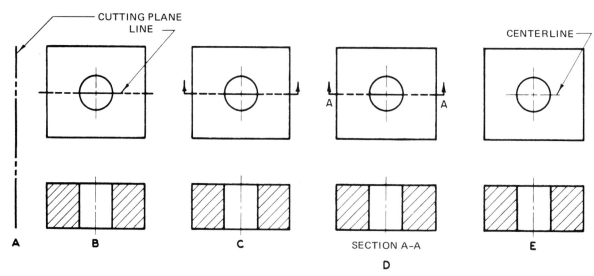

Fig. 2-2 Locating the cutting planes

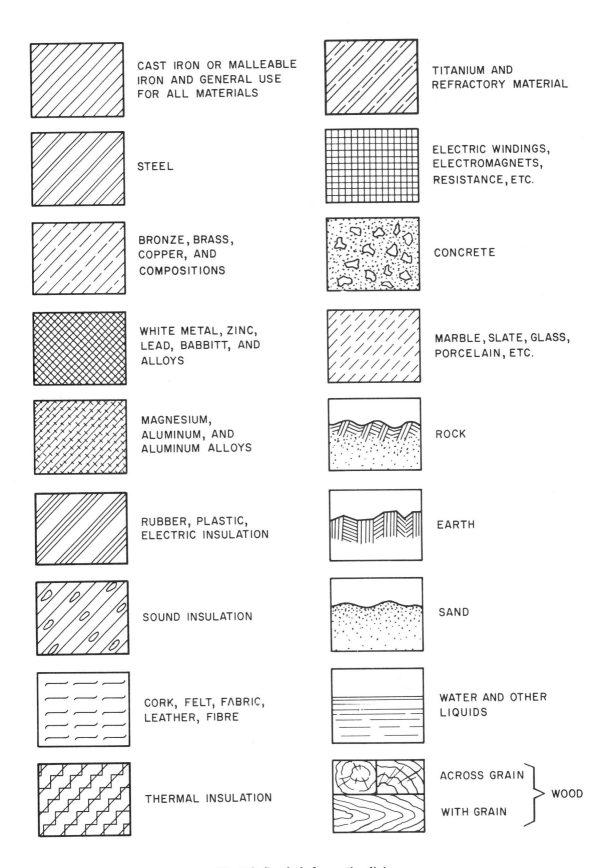

Fig. 2-3 Symbols for section lining

16 Unit 2 Section Views

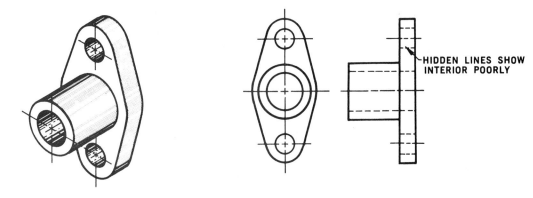

(A) SIDE VIEW NOT SECTIONED

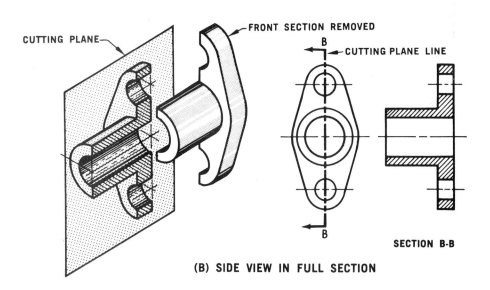

(B) SIDE VIEW IN FULL SECTION

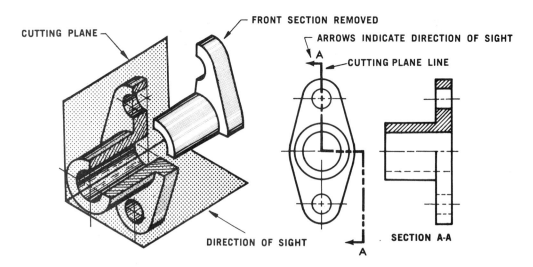

(C) SIDE VIEW IN HALF SECTION

Fig. 2-4 Full and half sections (from C. Jensen & R. Hines, *Interpreting Engineering Drawings — Metric Edition.* © 1979 by Delmar Publishers Inc.)

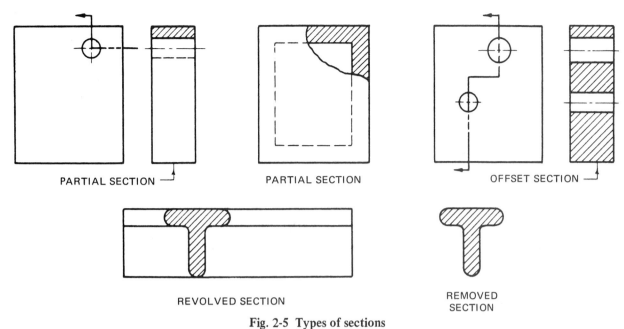

Fig. 2-5 Types of sections

All of these methods of designating cutting planes are acceptable. The drafter should select and use the one that clearly indicates where a section is taken. The mechanic in turn must be able to interpret the drawing and visualize this section.

OFFSET CUTTING PLANES

Offsetting or changing the straight line direction of the cutting plane is used to describe internal features of irregular objects. These features may not fall in a straight line and, therefore, the cutting plane line must be offset to pass through them, Figure 2-5.

SECTION LINES

The surfaces which have been cut through are indicated by a series of slant lines known as *section* or *crosshatch lines*. In drawing sections of various machine parts, slanted lines indicate the different materials of which the parts are made, Figure 2-3. Each material is represented by a different pattern of lines. On most detail drawings, however, sections are shown using the pattern for cast iron. The kind of material is then indicated in the specifications.

FULL SECTIONS

The type and number of sections depends on the complexity of the part. A *full section* is one in which an imaginary cut has been made all the way through the object. The cut section of the object is then represented in a separate view called a *section view*. Hidden lines representing surfaces behind the cutting plane are left out. This helps to keep the view clear for better understanding. When more than one section is required, the section view is identified with letters, Figure 2-4B.

HALF SECTIONS

A half section is often used for symmetrical objects. A *half section* shows a cutaway view of only one half of the part, Figure 2-4C. One advantage of half sections is that both an interior and exterior view are shown in the same view.

Construction details of an object are often shown by using revolved or removed section views. These sections may appear on the conventional view or adjacent to it, Figure 2-5.

REVOLVED SECTIONS

Revolved sections are used to show the cross-sectional shape of spokes, ribs, cast arms, rods, or structural steel shapes. The view is obtained by passing a cutting plane through the object perpendicular to the centerline or axis of the part. The section is then rotated 90° on the view to reflect the true size and shape of the area in section. Visible lines on either side of the section are removed or break lines are added to isolate the view, Figure 2-6. Cutting plane lines are usually omitted on revolved section views.

REMOVED SECTIONS

Removed sections are very similar to revolved sections. However, a removed section is taken out of the normal projection position in relation to standard views, Figure 2-7. A removed section is usually placed on the drawing in a convenient place and labeled section A-A, section B-B, and so on. The letters identifying the section should correspond with the letters at the end of the cutting plane lines.

A removed section often is a partial section view. Frequently, the view is enlarged to permit greater clarification, as shown on the Spider drawing, D-16.

BROKEN-OUT SECTIONS

When only a partial section view is needed a broken-out section may be used. A broken-out section isolates one area of the object for internal clarification. A heavy break line is used to define the boundaries of the view, Figure 2-8. Broken-out sections are used where less than a half section is needed.

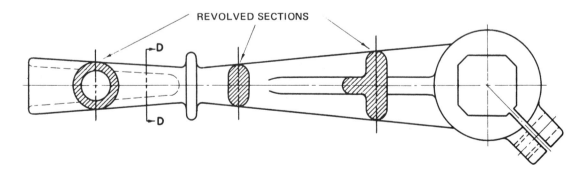

Fig. 2-6 Revolved sections

Unit 2 Section Views 19

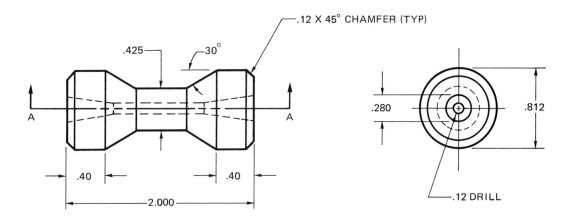

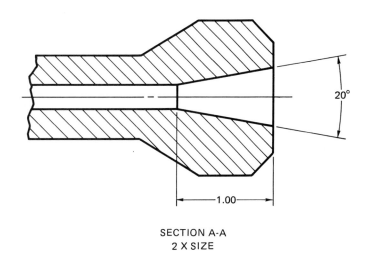

Fig. 2-7 Removed section

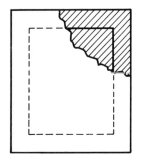

Fig. 2-8 Broken-out section

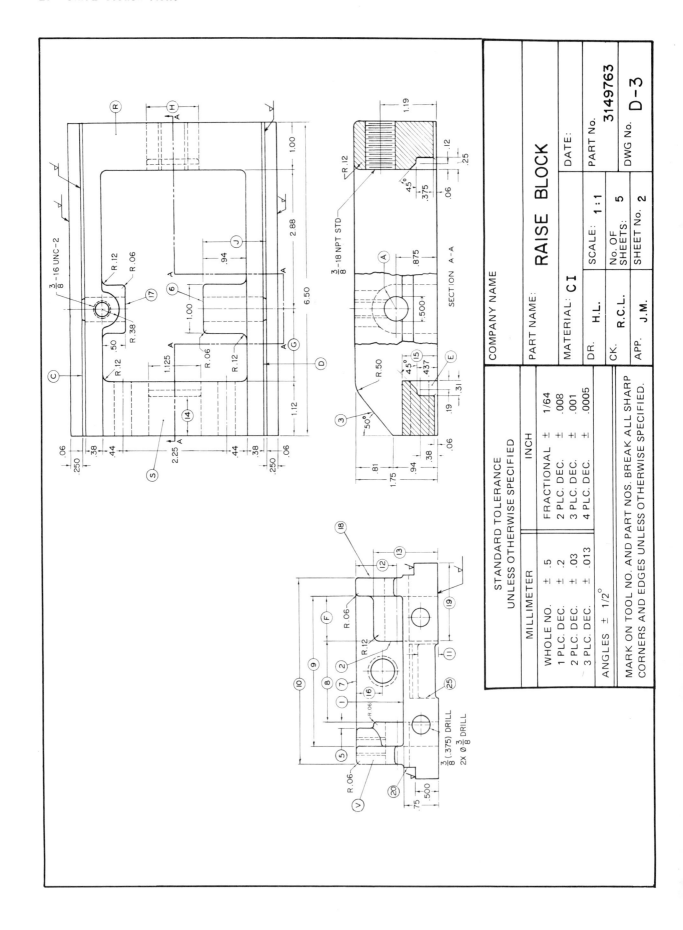

ASSIGNMENT D-3: RAISE BLOCK

1. What type of line is used to show where section A-A is taken? _____
2. What type of material is indicated by the section lines? _____
3. Give the number and the size of the small drilled holes. _____
4. What surface in the top view does ⑦ represent? _____
5. What surface in the top view does ① represent? _____
6. What line or surface in the left view does ⑥ represent? _____
7. What line or surface in the front view does Ⓥ represent? _____
8. What type of section is shown? _____
9. Determine distance ⑤ . _____
10. Determine distance ⑧ . _____
11. Determine distance ⑨ . _____
12. Determine distance ⑩ . _____
13. Determine distance ⑪ . _____
14. Determine distance ⑫ . _____
15. What is dimension ⑬ ? _____
16. What is dimension ⑭ ? _____
17. What is dimension ⑮ ? _____
18. What line or surface in the top view does Ⓑ represent? _____
19. Determine distance ⑯ . _____
20. Surface Ⓔ is represented by a line in the left view. Indicate the line. _____
21. What line or surface in the left view is represented by Ⓓ ? _____
22. Locate in the left view the line or surface that is represented by line Ⓒ . _____
23. How many 3/8 (.375) drilled holes are required? _____
24. Determine distance ⑲ . _____
25. Determine the overall width of the left view. _____
26. Determine radius Ⓐ . _____
27. Determine distance Ⓕ . _____
28. Determine distance Ⓖ . _____
29. Determine distance Ⓗ . _____
30. Determine distance Ⓙ . _____

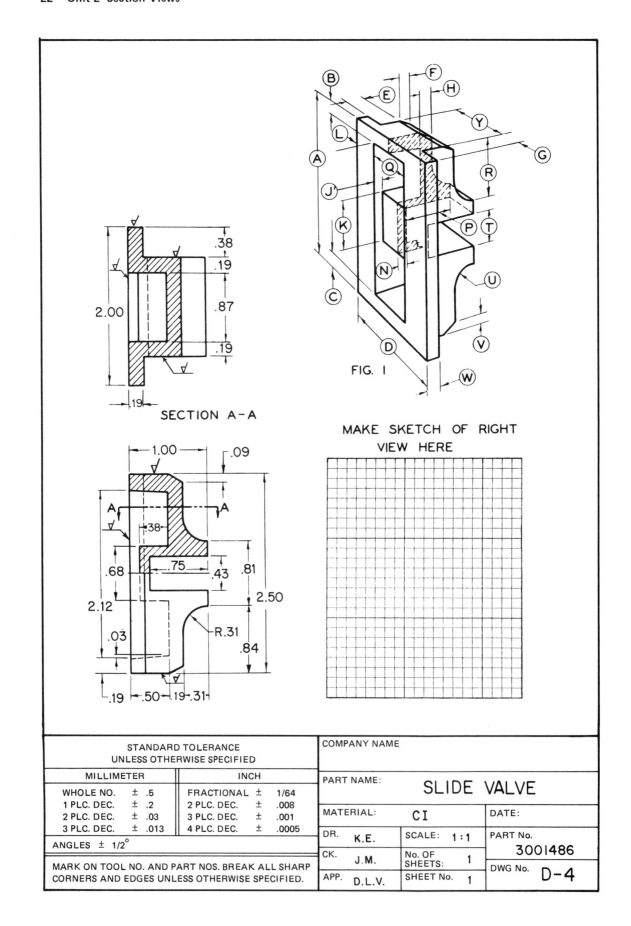

ASSIGNMENT D-4: SLIDE VALVE

1. Make a freehand sketch showing the right view of the Slide Valve in the space provided on the drawing.

2. Determine the distances or dimensions for each of the following letters found on the drawing of the Slide Valve.

   Ⓐ   Ⓕ   Ⓛ   Ⓣ
   Ⓑ   Ⓖ   Ⓝ   Ⓤ
   Ⓒ   Ⓗ   Ⓟ   Ⓥ
   Ⓓ   Ⓙ   Ⓠ   Ⓦ
   Ⓔ   Ⓚ   Ⓡ   Ⓨ

Ⓐ = _____
Ⓑ = _____
Ⓒ = _____
Ⓓ = _____
Ⓔ = _____
Ⓕ = _____
Ⓖ = _____
Ⓗ = _____
Ⓙ = _____
Ⓚ = _____
Ⓛ = _____
Ⓝ = _____
Ⓟ = _____
Ⓠ = _____
Ⓡ = _____
Ⓣ = _____
Ⓤ = _____
Ⓥ = _____
Ⓦ = _____
Ⓨ = _____

unit 3

SURFACE TEXTURE

Surface texture refers to the degree of quality required on the surface of a workpiece. Modern technology demands close tolerances, high speeds, and increased resistance to friction and wear. To accomplish this, exact control of surface texture must be maintained. Simple finish marks are no longer adequate in all cases. Where specific texture quality must be controlled, special symbols are used.

SURFACE TEXTURE TERMINOLOGY

The American National Standards Institute (ANSI) recommends the use of standard symbols for surface texture. These symbols describe the allowable roughness, waviness, and height. Certain other terms are also used to describe surface characteristics. The following is a list of terms and definitions (also refer to Figure 3-1).

Roughness — High and low points on a surface. These are often caused by the machining process used to generate the surface.

Lay — Refers to the predominant direction of surface roughness caused by the machining process.

Waviness — The larger undulations of a surface which lie below the surface roughness marks. Roughness and lay characteristics are imposed on top of surface waviness.

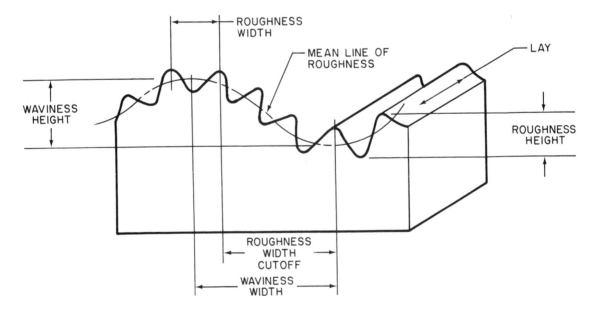

Fig. 3-1 Surface texture terminology

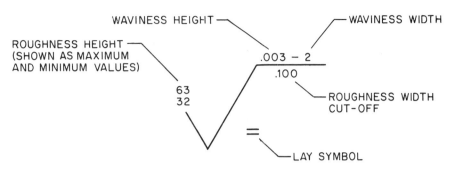

Fig. 3-2 Surface texture symbol

Microinch — A measurement in millionths of an inch. The height of surface roughness is measured in microinches. The higher the number of microinches, the rougher the surface.

Roughness height — An arithmetical average height as measured from the mean line of the roughness profile. The mean line is a point half way between a peak and valley. Roughness height is the amount of deviation from that mean line.

Roughness width — The distance between a point on a ridge to an equal point on the next ridge.

Waviness width — A distance measured in the same way as roughness width.

Waviness height — Distance between the mean roughness line measured at the top and bottom of the wave.

Roughness width cutoff — The distance of surface roughness to be included in calculating average roughness height.

SURFACE TEXTURE SYMBOLS

To specify a surface quality, a special surface texture symbol is used. The symbol appears as a check mark with a horizontal bar across the top. Numerical values placed around the symbol specify allowable tolerances for surface texture, Figure 3-2.

Surface texture symbols, like dimensions, should only appear once on a drawing. They should not be used in more than one view to represent the same surface. When placed on a drawing, the symbol always is shown in an upright position, Figure 3-3.

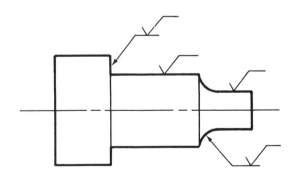

Fig. 3-3 Surface texture symbols are always in an upright position

SYMBOL	DESIGNATION	EXAMPLE
=	Lay parallel to the line representing the surface to which the symbol is applied	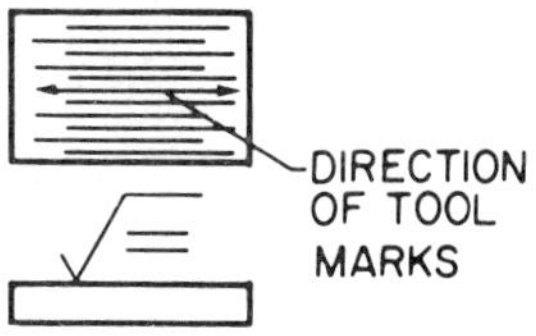 DIRECTION OF TOOL MARKS
⊥	Lay perpendicular to the line representing the surface to which the symbol is applied	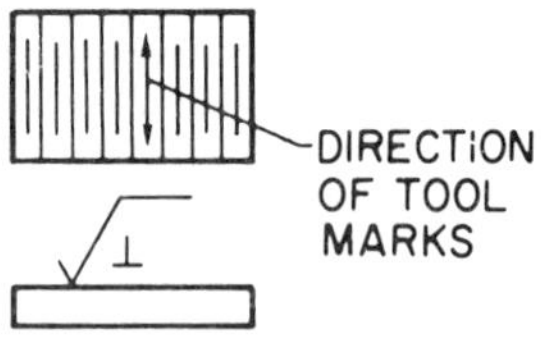 DIRECTION OF TOOL MARKS
X	Lay angular in both directions to line representing the surface to which the symbol is applied	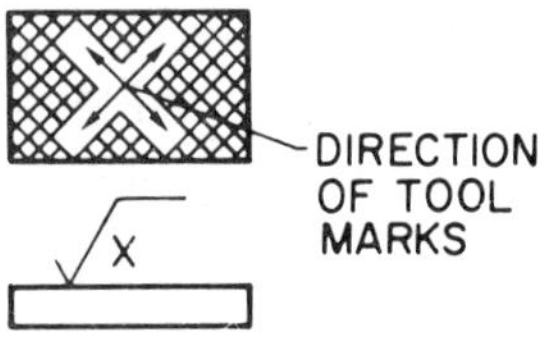 DIRECTION OF TOOL MARKS
M	Lay multidirectional	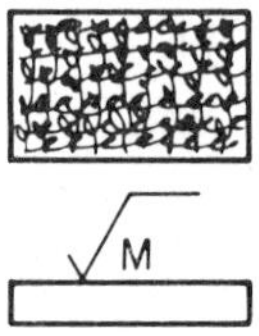
C	Lay approximately circular relative to the center of the surface to which the symbol is applied	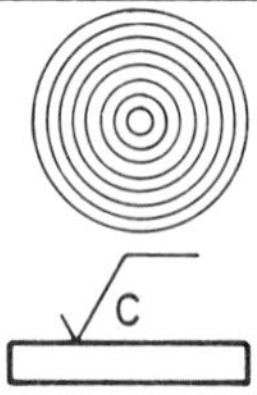
R	Lay approximately radial relative to the center of the surface to which the symbol is applied	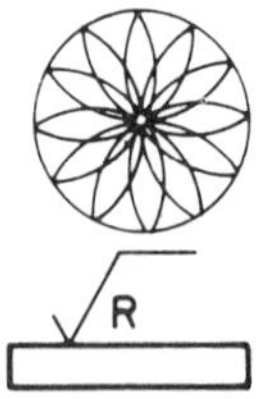
P	Lay nondirectional, pitted, or protuberant	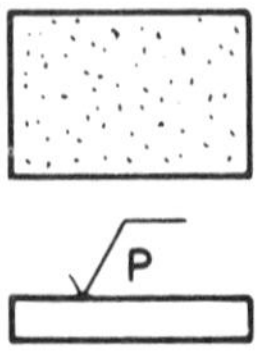

Fig. 3-4 Lay symbols (from C. Jensen & R. Hines, *Interpreting Engineering Drawings – Metric Edition.* © 1979 by Delmar Publishers Inc.)

LAY SYMBOLS

Lay symbols are used to represent the direction of the tool marks on a specified surface. Lay is specified for both function and appearance. Wear, friction, and lubricating qualities may be affected by lay. Figure 3-4 shows various lay symbols and how they might appear on the workpiece.

MEASURING SURFACE TEXTURE

The surface characteristic which is most often regarded as critical is roughness height. The instrument used to measure roughness height is called a *profilometer*. A profilometer has a stylus or surface indicator tip which reads surface roughness. As the stylus is moved across the surface, readings are displayed on a meter. The meter displays a numerical figure which is the arithmetical average of roughness height, Figure 3-5.

Surface roughness readings are given in microinches. A microinch is one millionth of an inch. The higher the reading in microinches, the rougher the surface finish. Microinch readings may be represented by the symbol (μ in). Figure 3-6 shows typical applications and corresponding microinch ratings.

Table 3-1 gives the ranges of surface roughness normally specified for selected manufacturing processes.

Surface control should only be applied to drawings where it is essential to the part. If the fit or function will be affected, then surface texture may need definition. The unnecessary control of texture may lead to increased production costs.

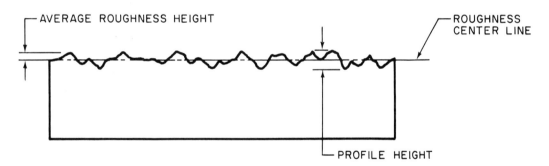

Fig. 3-5 Reading surface roughness

MICROINCHES AA RATING	APPLICATION
1000	Rough, low grade surface resulting from sand casting, torch or saw cutting, chipping, or rough forging. Machine operations are not required because appearance is not objectionable. This surface, rarely specified, is suitable for unmachined clearance areas on rough construction items.
500	Rough, low grade surface resulting from heavy cuts and coarse feeds in milling, turning, shaping, boring, and rough filing, disc grinding and snagging. It is suitable for clearance areas on machinery, jigs, and fixtures. Sand casting or rough forging produces this surface.
250	Coarse production surfaces, for unimportant clearance and cleanup operations, resulting from coarse surface grind, rough file, disc grind, rapid feeds in turning, milling, shaping, drilling, boring, grinding, etc., where tool marks are not objectionable. The natural surfaces of forgings, permanent mold castings, extrusions, and rolled surfaces also produce this roughness. It can be produced economically and is used on parts where stress requirements, appearance, and conditions of operations and design permit.
125	The roughest surface recommended for parts subject to loads, vibration, and high stress. It is also permitted for bearing surfaces when motion is slow and loads light or infrequent. It is a medium commercial machine finish produced by relatively high speeds and fine feeds taking light cuts with sharp tools. It may be economically produced on lathes, milling machines, shapers, grinders, etc., or on permanent mold castings, die castings, extrusion, and rolled surfaces.
63	A good machine finish produced under controlled conditions using relatively high speeds and fine feeds to take light cuts with sharp cutters. It may be specified for close fits and used for all stressed parts, except fast rotating shafts, axles, and parts subject to severe vibration or extreme tension. It is satisfactory for bearing surfaces when motion is slow and loads light or infrequent. It may also be obtained on extrusions, rolled surfaces, die castings and permanent mold castings when rigidly controlled.
32	A high-grade machine finish requiring close control when produced by lathes, shapers, milling machines, etc., but relatively easy to produce by centerless, cylindrical, or surface grinders. Also, extruding, rolling or die casting may produce a comparable surface when rigidly controlled. This surface may be specified in parts where stress concentration is present. It is used for bearings when motion is not continuous and loads are light. When finer finishes are specified, production costs rise rapidly; therefore, such finishes must be analyzed carefully.
16	A high quality surface produced by fine cylindrical grinding, emery buffing, coarse honing, or lapping, it is specified where smoothness is of primary importance, such as rapidly rotating shaft bearings, heavily loaded bearing and extreme tension members.
8	A fine surface produced by honing, lapping, or buffing. It is specified where packings and rings must slide across the direction of the surface grain, maintaining or withstanding pressures, or for interior honed surfaces of hydraulic cylinders. It may also be required in precision gauges and instrument work, or sensitive value surfaces, or on rapidly rotating shafts and on bearings where lubrication is not dependable.
4	A costly refined surface produced by honing, lapping, and buffing. It is specified only when the design requirements make it mandatory. It is required in instrument work, gauge work, and where packing and rings must slide across the direction of surface grain such as on chrome plated piston rods, etc. where lubrication is not dependable.
2, 1	Costly refined surfaces produced only by the finest of modern honing, lapping, buffing, and superfinishing equipment. These surfaces may have a satin or highly polished appearance depending on the finishing operation and material. These surfaces are specified only when design requirements make it mandatory. They are specified on fine or sensitive instrument parts or other laboratory items, and certain gauge surfaces, such as precision gauge blocks.

Fig. 3-6 Microinch ratings and typical applications (from C. Jensen & R. Hines, *Interpreting Engineering Drawings – Metric Edition.* © 1979 by Delmar Publishers Inc.)

ASSIGNMENT D-5: SHAFT INTERMEDIATE SUPPORT

1. Determine distance Ⓐ . _____

2. What is the surface finish requirement on the .500 diameter hole? _____

3. What does the lay symbol on the .390 diameter hole indicate? _____

4. Determine distance Ⓑ . _____

5. How many 25/64 holes are to be drilled? _____

6. Determine the left to right distance between the .390 hole and the .500 hole. _____

7. What two views of the Shaft Intermediate Support are shown? _____

8. What surface finish is required on the bottom surface Ⓔ of the Support? _____

9. What does the lay symbol indicate on the bottom surface Ⓔ callout? _____

10. Determine distance Ⓒ . _____

11. What instrument is used to measure roughness height? _____

12. Determine distance Ⓓ . _____

13. Determine the overall width of the Support from left to right in the right-side view. _____

14. Determine the overall height of the Support from top to bottom. _____

15. Sketch the bottom view in the space provided on the drawing.

unit 4

VIOLATIONS OF TRUE PROJECTION

The cost of a machine part includes three important factors. To be considered are design cost, material cost, and construction cost. Machine designers must be knowledgable in material specifications and applications. They must also strive to reduce both design time and construction time.

Frequently a designer will violate some basic projection principles to reduce time or to simplify a view. Two examples of such violations are rotated projection and untrue projection.

ROTATED OR ALIGNED PROJECTION

In Figure 4-1A, an object is drawn in true projection with an auxiliary view. In Figure 4-1B, the same information is provided in two views using the rotated or aligned method of projection. This method is preferred because it is less confusing and reduces drawing time.

In Figure 4-2, two section views, M-M and M-N, are provided. The section taken at M-M is shown in true projection at B. Section M-N is shown at C using the rotated or aligned method of projection.

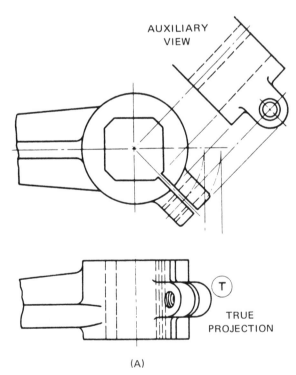

Fig. 4-1 Rotated or aligned projection

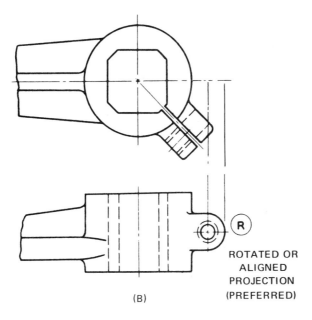

(B)

ROTATED OR ALIGNED PROJECTION (PREFERRED)

Fig. 4-1 Continued

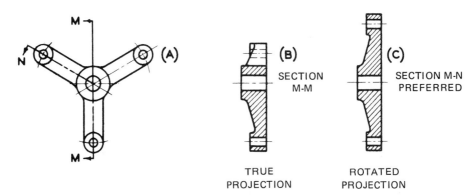

Fig. 4-2 Section views

UNTRUE PROJECTION

Frequently the edges of different surfaces are on the same plane. These surfaces, when projected to another view, will appear as a single line. This may lead to some confusion in print interpretation. Therefore, it is common practice to move a line out of its true projected position. This provides added clarity to the view.

In Figure 4-3A, the top of the hole and the bottom of the slot are on the same plane. When projected into a side view each appears as a single line. At B, the top of the hole has been moved out of its true projection position in the side view. This method provides a distinction between the hole and slot surfaces. Although this practice is a projection violation, it brings out an important detail of the part.

An additional example is provided in Figure 4-4. The object shown at A is drawn in true projection. However, the untrue projection at B is easier to draw and is clearly understood.

34   Unit 4   Violations of True Projection

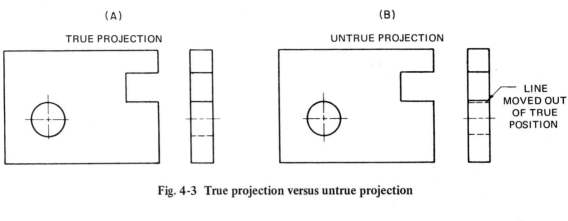

Fig. 4-3 True projection versus untrue projection

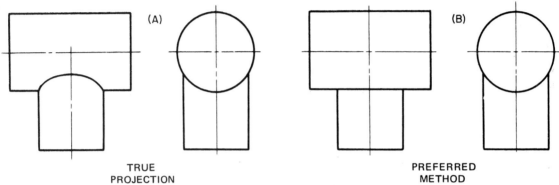

Fig. 4-4 Untrue projection

The block shown in Figure 4-5 is made with a round hole on one face and a square hole on the other. To clarify the drawing, two side views are used without any hidden lines shown.

UNTRUE PROJECTION OF SECTIONS

The partitions or webs and spokes of a part are not usually shown in cross section. While this is considered a violation of true projection, it is preferred practice due to drawing ease and clarity.

The object illustrated in Figure 4-6 shows three webs or ribs connected to a center hub. Section P-P is taken through the center hole and splits the upper web. The section view at A illustrates the true projection of P-P. However, the more conventional and otherwise preferred method is shown at B.

A similar application used in sectioning the spokes of a pulley is shown in Figure 4-7. The true section C-C is shown at A. The section view at B is preferred because it eliminates the section lining of the spokes. The assumption is that the view of the arms is taken at D-D.

Fig. 4-5

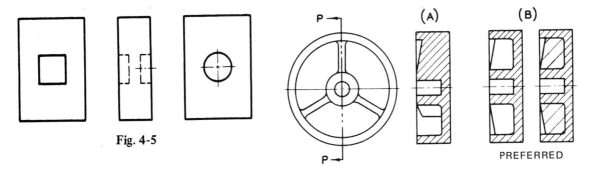

Fig. 4-6 Sectioning of webs and spokes

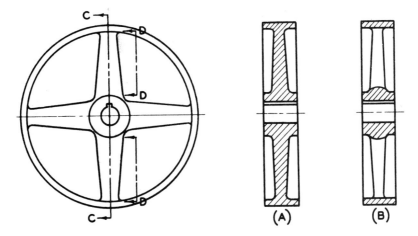

Fig. 4-7 Sectioning the spokes of a pulley

Unit 4 Violations of True Projection

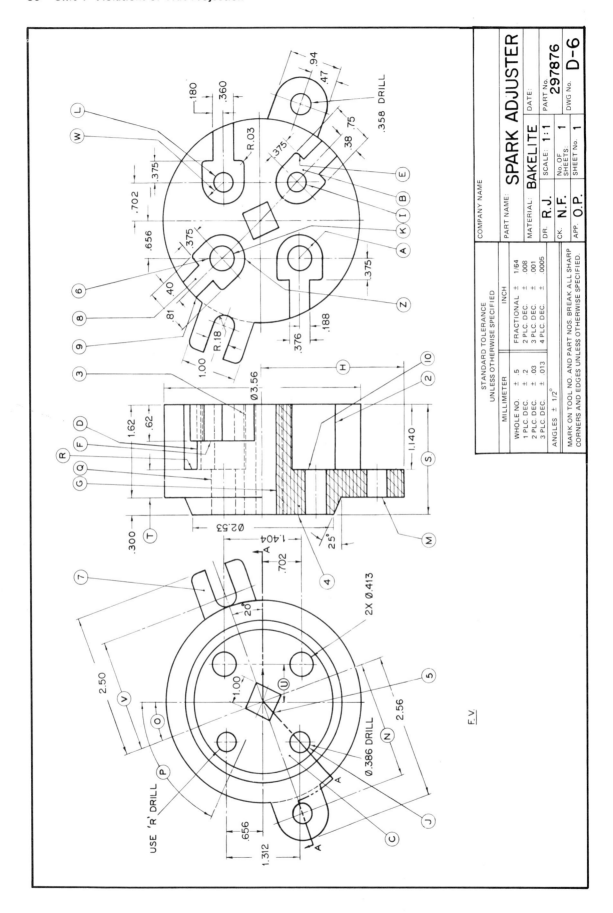

ASSIGNMENT D-7: COIL FRAME

1. What part or detail of the front view do the lines at ② represent?
2. How is distance ③ determined?
3. In how many places on the front view is lug ④ shown?
4. What line in the front view does the line at ⑤ represent?
5. Locate point ⑧ on the front view.
6. Locate line ⑭ on the right side view.
7. Locate surface ⑮ on the front view.
8. What line or point in the right view represents circle ⑯?
9. What line or surface in the right view represents the surface at ⑰?
10. What line or surface in the right view is represented by line ⑱?
11. What line in the right view represents surface ㉒?
12. What surface in the front view does line ㉔ represent?
13. What is distance Ⓝ?
14. Determine the angular measurement at Ⓥ.
15. Determine the angular measurement at Ⓦ.
16. What is distance Ⓐ?
17. Determine radius Ⓑ.
18. What is distance Ⓒ?
19. What is distance Ⓓ?
20. What is distance Ⓔ?
21. Give the thickness of lug Ⓕ.
22. Determine radius Ⓙ.
23. Determine radius Ⓚ.
24. Determine the following distances:

    Ⓖ   Ⓞ   Ⓣ
    Ⓗ   Ⓟ   Ⓤ
    Ⓘ   Ⓠ   Ⓛ
    Ⓧ   Ⓡ   Ⓜ
    Ⓨ   Ⓢ

Ⓖ = _____
Ⓗ = _____
Ⓘ = _____
Ⓧ = _____
Ⓨ = _____
Ⓞ = _____
Ⓟ = _____
Ⓠ = _____
Ⓡ = _____
Ⓢ = _____
Ⓣ = _____
Ⓤ = _____
Ⓛ = _____
Ⓜ = _____

unit 5

SPECIAL VIEWS

Special views are sometimes needed to clarify a drawing and make it easier to interpret. Parts with complex detail often require extra views which aid the print reader.

PARTIAL VIEWS

A partial view is one in which only part of an object is shown. Sufficient information should be given in the partial view to complete the description of the object. Figure 5-1 illustrates an example of proper use of a partial view. A break line is provided to show where the view ends.

Partial views provide the following advantages:

- They save drawing time.

- They conserve space which might otherwise be required to draw the object completely.

- They sometimes permit a larger scale drawing. This allows details to be brought out more clearly.

Symmetrical objects may be shown in half views. The partial view is drawn on one side of the centerline as shown in Figure 5-2.

In the case of the Coil Frame (D-7), a partial view is used. This allows the object to be drawn to a larger scale. The larger view provides clarity and saves drawing time and drawing space.

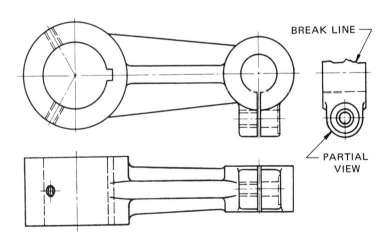

Fig. 5-1 Partial view

42 Unit 5 Special Views

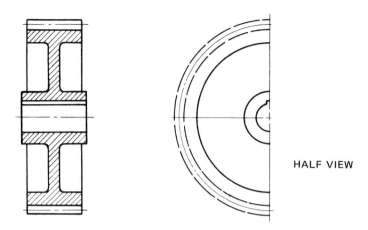

Fig. 5-2 Half view of a symmetrical object

DISTORTED VIEWS

A distorted view is also used where additional clarity is required. Although it may be considered a violation of true projection, it simplifies complex symmetrical parts.

Figure 5-3 illustrates a true projection view of the wheel. The distorted view shown in Figure 5-4 is easier to interpret and is preferred.

The half section D-D required in Figure 5-4 would be shown the same as B in Figure 2-3.

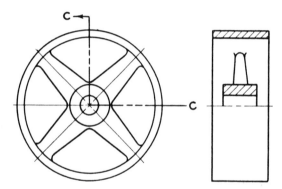

Fig. 5-3 True projection

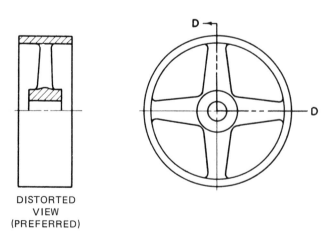

Fig. 5-4 Distorted view

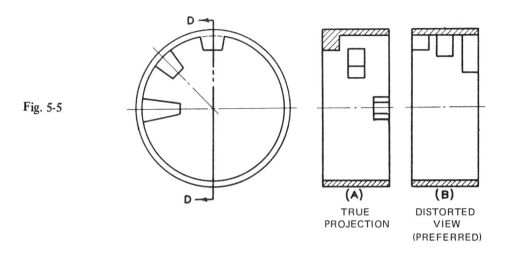

Fig. 5-5

(A) TRUE PROJECTION
(B) DISTORTED VIEW (PREFERRED)

Another example of using a distorted view is shown in Figure 5-5. A ring with three lugs cast on the inside of the rim is shown at A in true projection. The distorted view at B is much clearer and therefore preferred.

The drawing of the Coil Frame (D-7) illustrates the use of distorted views. The information concerning the tapped lugs, stops, and T-slots is best shown in section A-A by distorting and rotating the view.

BOTTOM VIEWS

Bottom views are not ordinarily used in standard drawing practice and thus may be considered a special view. Prints with top, front, and right-side views, or some combination of these views, are most common.

Bottom views are used when it is necessary to bring out details not clearly shown in other views. The bottom view should include only those lines which are necessary to complete the description of the object, Figure 5-6. By the same rule, the lines shown on the bottom view may be omitted from the top view.

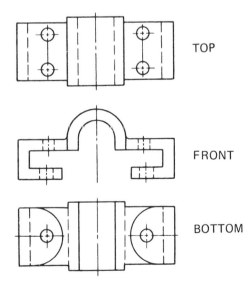

Fig. 5-6 Application of the bottom view

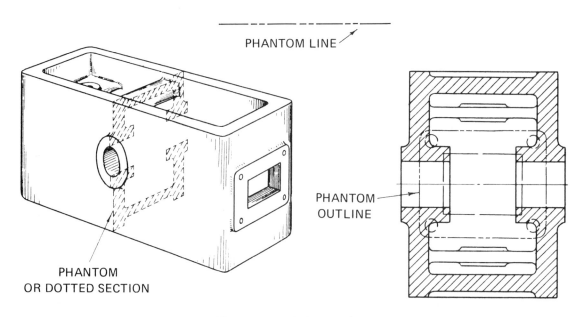

Fig. 5-7 Use of phantom lines

PHANTOM LINES AND VIEWS

Phantom lines are long dash lines used to indicate the position of an absent part in relation to the view which is shown. Phantom lines or views drawn in phantom clarify the drawing without the need for additional views.

Figure 5-7 shows how phantom lines and views are used to indicate the position of a cover plate mounting hole. To aid in the interpretation of the object, both a phantom section of the part and a phantom outline of the plate surface are shown.

Phantom lines are also used to show parts of an object which may later be removed. This is often the case when machining lugs are required. Machining lugs may be provided on an object to allow additional clamping to secure the part during machining. In each corner of the Drive Housing (Unit 11), rectangles are shown in phantom representing lugs molded on the casting.

Another example of a machining lug is shown in Figure 5-8. This lug is used to provide a flat surface on the end of the object for centering. It also gives a uniform center bearing for additional machining operations. The lug will later be removed if it interferes with the operation of the part.

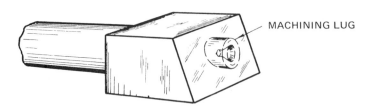

Fig. 5-8 Application of a machining lug

Unit 5 Special Views 45

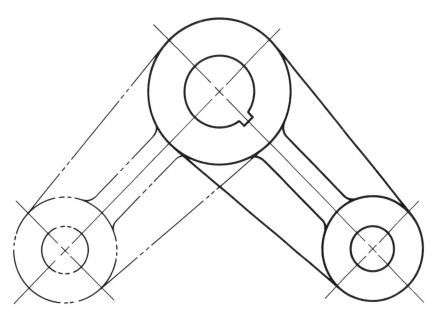

Fig. 5-9 Alternate positions

Phantom views are also used to show mechanisms in alternate positions or how they connect with adjacent parts, Figure 5-9. Phantom lines clarify the operation of the mechanism even though the view drawn in phantom may not be an actual part of the detail or assembly. As an example, refer to the track which the Four-Wheel Trolley rides on (Unit 16 Assignment drawing).

Additional uses of phantom lining include applications where repetitive detail is required on a drawing. Rather than spend time by drawing identical features, the drafter may use phantom lines to show the continuation of a feature. This method is often applied to long threaded shafts, springs, or gears, Figure 5-10.

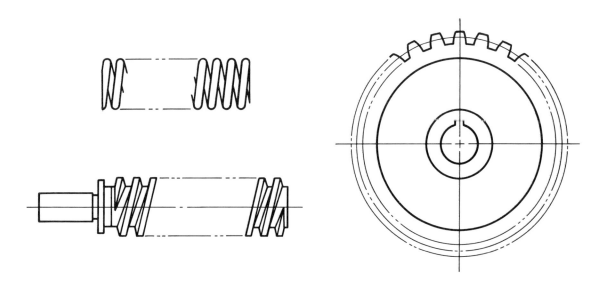

Fig. 5-10 Use of phantom lines to show repetitive features

ASSIGNMENT D-8: INDEX PEDESTAL

1. What is the diameter at Ⓔ ? _____
2. What line does ⑦ represent in the front view? _____
3. What is dimension at Ⓣ ? _____
4. What is the diameter of circle Ⓢ ? _____
5. What is distance Ⓓ ? _____
6. Determine distance Ⓕ . _____
7. How deep is the square hole Ⓘ ? _____
8. Determine distance Ⓚ . _____
9. From what point in the bottom view is line ⑥ projected? _____
10. Determine distance Ⓞ . _____
11. Determine distance Ⓙ . _____
12. What measurement would be used for distance Ⓝ ? _____
13. What measurement would be used for distance Ⓜ ? _____
14. Determine distance Ⓗ . _____
15. Determine distance Ⓒ . _____
16. Determine distance Ⓖ . _____
17. Determine the size of hole Ⓑ . _____
18. Surface at Ⓥ is represented by a line or surface in the front view. Which one is it? _____
19. Locate line ④ in the right view. _____
20. Determine distance Ⓧ . _____
21. Determine the overall height of the Pedestal. _____
22. Which line or surface represents surface Ⓨ in the right view? _____
23. Determine distance Ⓐ . _____
24. Determine distance Ⓛ . _____
25. Determine distance Ⓠ . _____

48 Unit 5 Special Views

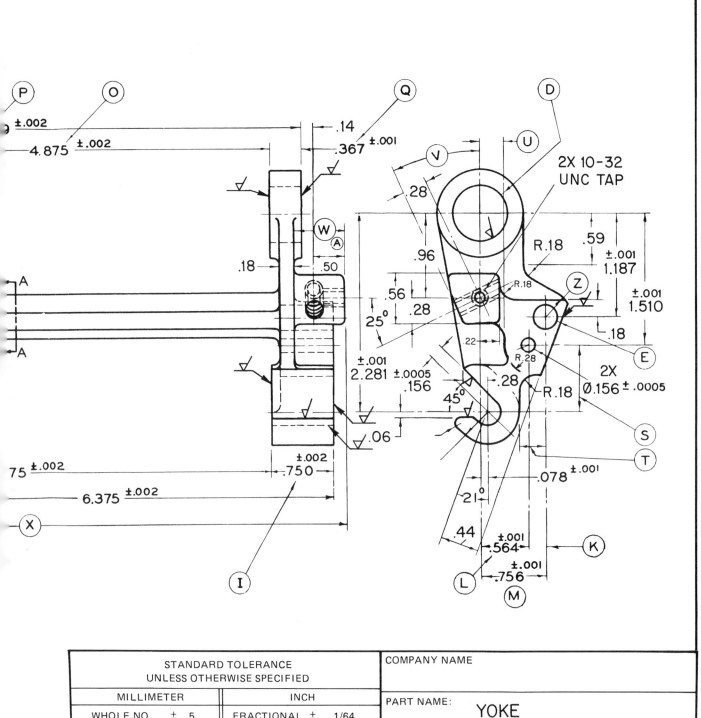

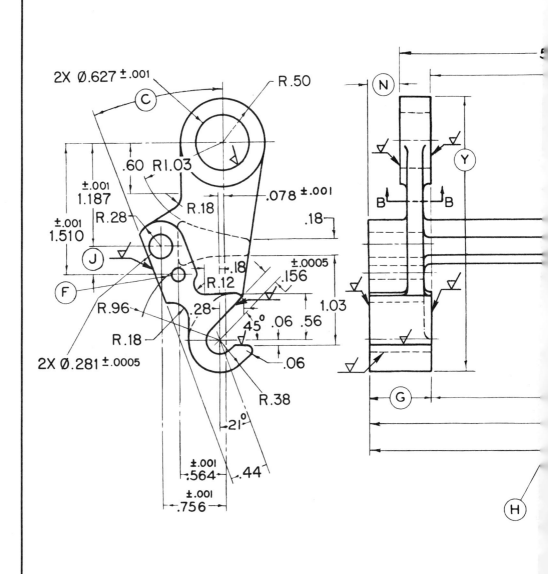

ASSIGNMENT D-9: YOKE

1. Make a freehand sketch of a section on A-A, approximately to scale, in the grid provided to the right.

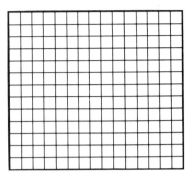

2. Make a freehand sketch of a section on B-B, approximately full size, assuming all corners to have a .06″ radius. Dimension where necessary. Place the sketch in the grid provided to the right.

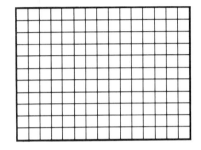

3. Determine angle Ⓒ . _____
4. What is the size of hole Ⓓ ? _____
5. What is the size of hole Ⓔ ? _____
6. What is the size of hole Ⓕ ? _____
7. What is distance Ⓖ ? _____
8. If dimension Ⓗ is made exactly 4.877, what will the high limit be for Ⓖ and Ⓘ ? _____
9. What is distance Ⓙ ? _____
10. If dimension Ⓛ is made exactly .564 and dimension Ⓜ exactly .757, what will dimension Ⓚ have to be? _____
11. Determine distance Ⓝ if distances Ⓖ , Ⓞ , Ⓟ , and Ⓠ are made .001 over given sizes. _____
12. Determine distance Ⓢ . _____
13. Determine distance Ⓤ . _____
14. Determine distance Ⓣ . _____
15. Determine angle Ⓥ . _____
16. Determine distance Ⓦ . _____
17. Which circled letter indicates a changed dimension? _____
18. Determine the overall length Ⓧ . _____
19. Determine the overall height Ⓨ . _____
20. What is radius Ⓩ ? _____

unit 6

POSITIONAL DIMENSIONING

POINT-TO-POINT DIMENSIONS

Most linear (in line) dimensions apply on a point-to-point basis. Point-to-point dimensions are applied directly from one feature to another, Figure 6-1. Such dimensions are intended to locate surfaces and features directly between the points indicated. They also locate corresponding points on the indicated surfaces.

For example, a diameter applies to all diameters of a cylindrical surface. It does not merely apply to the diameter at the end where the dimension is shown. A thickness applies to all opposing points on the surfaces.

DATUM DIMENSIONING

Datum dimensioning is a system where a number of dimensions are given from a common point, Figure 6-2. This common point is called a *data point*. More than one point may be used as a datum, Figure 6-3A.

Datum, or baseline dimensioning as it is sometimes called, is used in accurate layout work. A common data point helps overcome errors which may accumulate in the buildup of tolerances between point-to-point dimensions.

The dimensions for the curved surface in Figure 6-4 are taken from the datum line. The datum used in Figure 6-5 is a centerline. The tolerances from the datum points must be held to one-half the tolerance acceptable between features. In Figure 6-3B, if the tolerance between the holes is ± .002, the datum dimensions would be ±.001. Datum dimensions make it easier to read a drawing and insure greater accuracy in building the part.

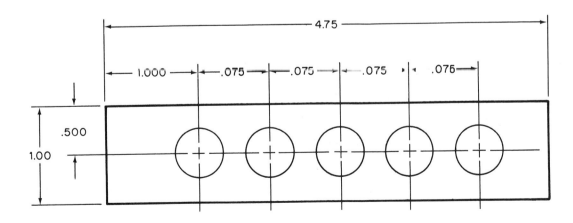

Fig. 6-1 Point-to-point dimensioning

52   Unit 6 Positional Dimensioning

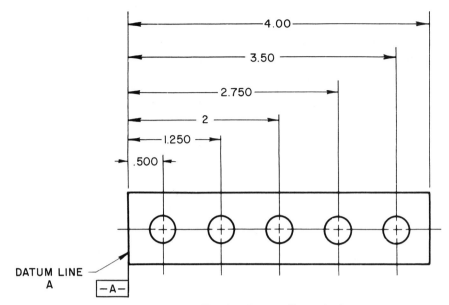

Fig. 6-2 Datum dimensioning

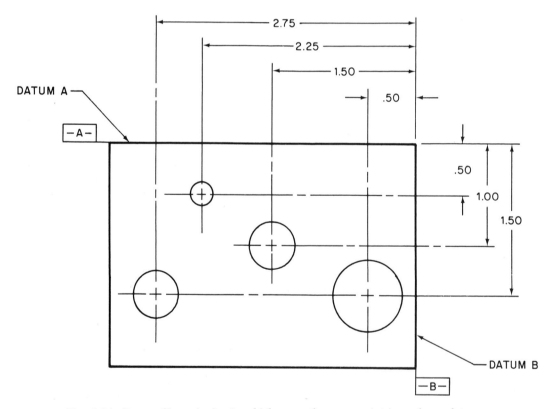

Fig. 6-3A Datum dimensioning in which more than one point is used as a datum

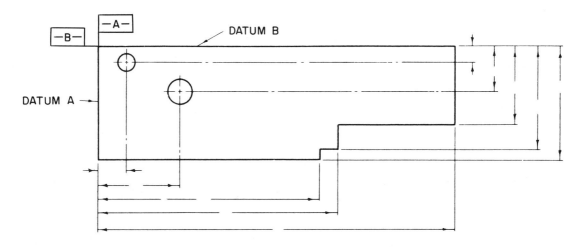

Fig. 6-3B Datum dimensioning from machined edges

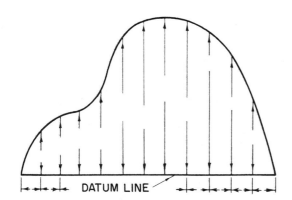

Fig. 6-4 Datum line for a curved surface (from C. Jensen & R. Hines, *Interpreting Engineering Drawings.* © 1970 by Delmar Publishers Inc.)

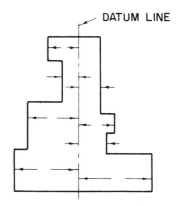

Fig. 6-5 The centerline is the datum line in this illustration (from C. Jensen & R. Hines, *Interpreting Engineering Drawings.* © 1970 by Delmar Publishers Inc.).

Unit 6 Positional Dimensioning

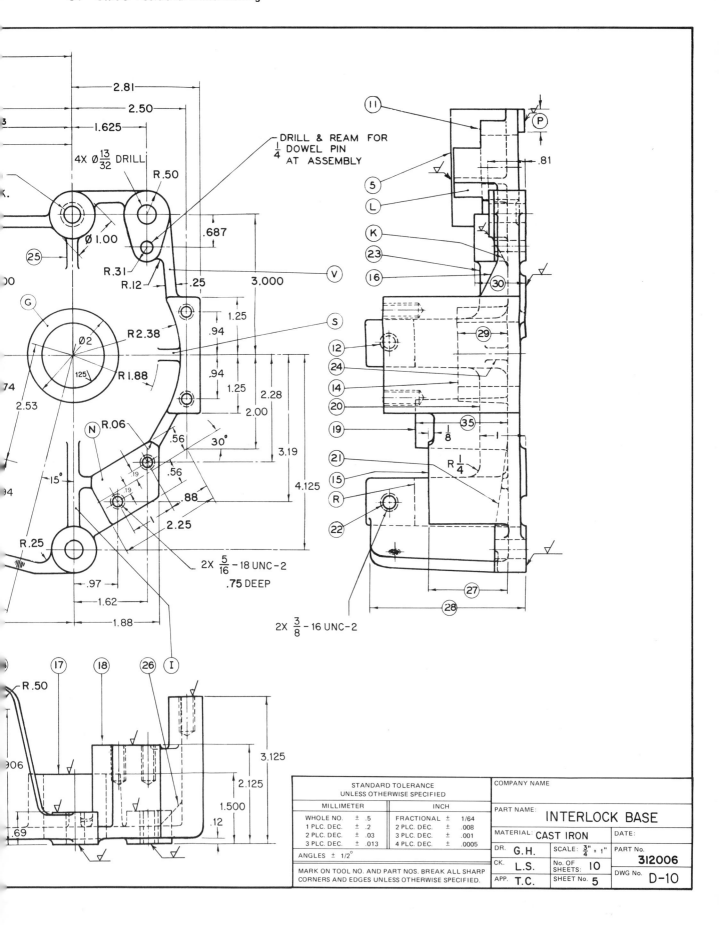

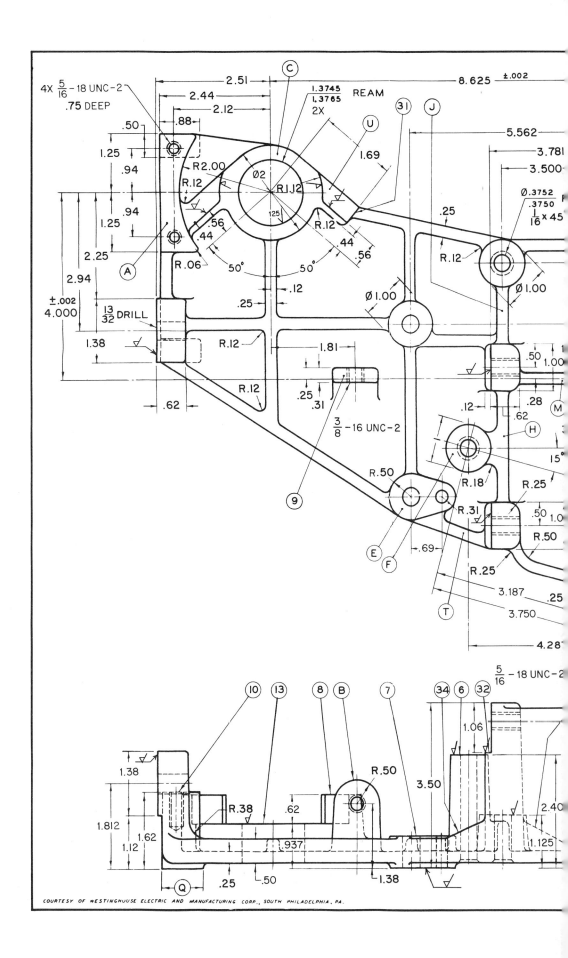

ASSIGNMENT D-10: INTERLOCK BASE

1. Locate surface Ⓐ in the front view and the right-side view.
2. Locate lug Ⓑ in the top view.
3. Find surface Ⓒ in the front view and the right-side view.
4. Locate surface Ⓔ in the front view.
5. Find surface Ⓕ in the front view and the right-side view.
6. Locate surface Ⓖ in the front view and the right-side view.
7. Find surface ⑱ in the top view and the right-side view.
8. What lines are used as datums for the dimensions in the front view?
9. What establishes the datum in the bottom view?
10. How are the $\frac{1.3745}{1.3765}$ holes produced?
11. Locate surface Ⓗ in the right-side view.
12. Locate surface Ⓘ in the right-side view.
13. How deep is threaded hole ㉒ ?
14. Locate rib ㉓ in the top view.
15. Find rib ㉔ in the front view and the top view.
16. Determine distance ㉗ .
17. Determine distance ㉘ .
18. Determine distance ㉙ .
19. Give the size of hole ⑫ .
20. Determine distance ㉟ .
21. Indicate rib ㉕ in the right-side view.
22. Locate surface Ⓛ in the top view.
23. Determine distance ㉚ .
24. Locate surface Ⓡ in the front view.
25. Determine distance Ⓠ .
26. Determine distance Ⓟ .
27. Indicate surface Ⓜ in the front view.
28. Locate rib ㉞ in the top view.
29. Locate surface ⑧ in the top view.
30. Locate surface Ⓥ in the right view.

Unit 6 Positional Dimensioning

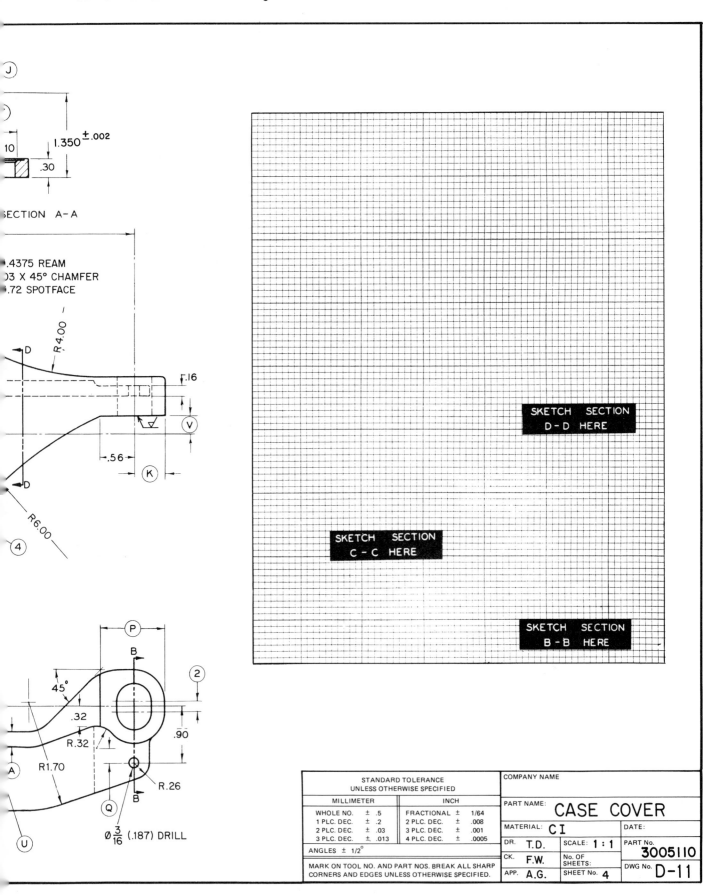

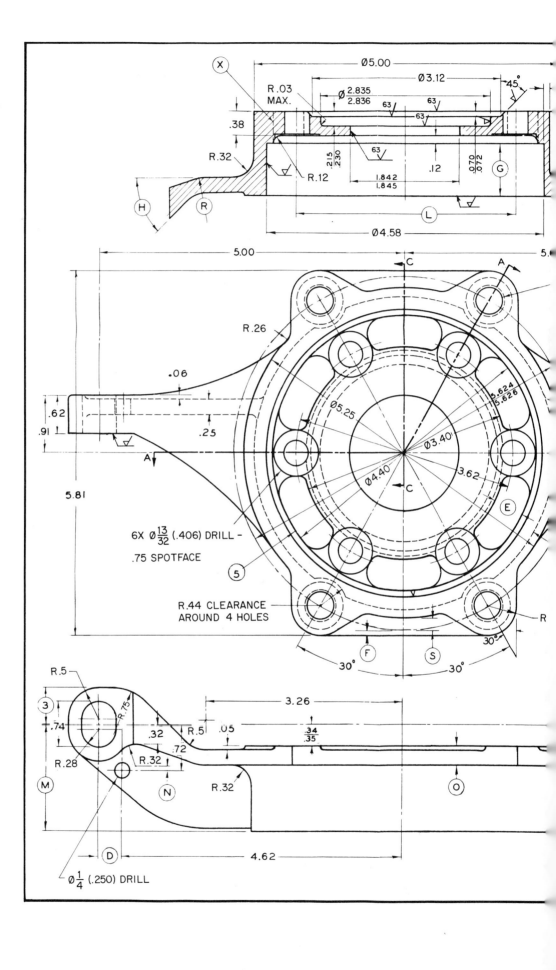

ASSIGNMENT D-11: CASE COVER

1. Sketch sections B-B, C-C, and D-D in the space provided for each section on the drawing of the Case Cover.

2. How many drilled and reamed holes are shown? _____

3. Determine distance ②. _____

4. Determine distance ③. _____

5. What surface finish is required on the 1.842/1.845 hole? _____

6. Determine distance Ⓐ. _____

7. Determine distance Ⓔ. _____

8. Determine distance Ⓖ. _____

9. Determine distance Ⓕ. _____

10. Determine distance Ⓓ. _____

11. Determine angle Ⓗ. _____

12. Determine distance Ⓙ. _____

13. Determine distance Ⓚ. _____

14. Determine distance Ⓛ. _____

15. Determine distance Ⓜ. _____

16. Determine distance Ⓝ. _____

17. Determine distance Ⓞ. _____

18. Determine distance Ⓟ. _____

19. Determine distance Ⓠ. _____

20. Determine radius Ⓡ. _____

21. Determine distance Ⓢ. _____

22. Determine diameter Ⓣ. _____

23. Locate ④ in the front view. _____

24. What line in section A-A indicates the surface represented by line ⑤? _____

25. Determine distance Ⓥ. _____

unit 7

GEOMETRIC TOLERANCES — DATUMS

Modern day manufacturing processes require precise tolerances to insure the interchangeability of parts. Mass-produced parts must be held within specified dimensional tolerances to achieve proper function and relationship to mating units. Geometric dimensioning controls the form or position of part features by means of a language of symbols. These symbols enable the print reader to interpret dimensional requirements and limit the amount of notes on a drawing. The geometric system of dimensioning is a widely accepted practice in industry.

This unit describes the key elements which apply to geometric dimensioning.

TERMINOLOGY

Allowance — The intentional difference in size between mating parts.

Basic Dimension — The exact theoretical dimension to which tolerances and allowances are applied.

Nominal Size — The stated designated size of an object which may or may not be the actual size.

Feature — The specific portion of an object to which dimensions and tolerances are applied. A feature may include one or more surfaces such as holes, slots, threads, etc.

Limits of Size — The applicable maximum and minimum size of a feature.

Form Tolerance — The amount of permissable surface variation from the basic or perfect form.

Positional Tolerance — The amount of permissable dimensional variation from basic or perfect location.

True Position — The term used to describe the perfect location of a point, line, or surface.

Datum — Points, lines, planes, cylinders, axes which are assumed to be exact for purposes of reference. Datums are established from actual features and are used to establish the relationship of other features.

Datum Axis — The theoretically exact centerline of a datum cylinder.

Datum Cylinder — (Or other geometrical form) The theoretically exact form profile of the actual datum feature surface.

Datum Feature — The actual part surface or feature used to establish a datum.

Datum Plane — The theoretically exact plane established by the extremities of the actual feature surface.

Specified Datum — A surface or feature identified with a datum symbol.

Maximum Material Condition — That condition which exists when a part feature contains the maximum amount of material.

Regardless of Feature Size — A tolerance of form or position which must be met regardless of where the feature lies within the size tolerance.

Unit 7 Geometric Tolerances — Datums

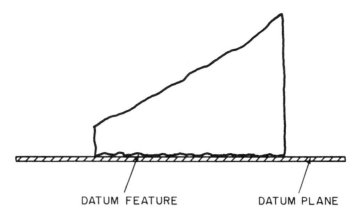

Fig. 7-1 Datum plane and datum feature

DATUMS

Datums are established by, or relative to, the actual features of a part. They are used as references from which other features are located. These datums may be points, lines, planes, axes, or cylinders. However, they must not be confused with datum features. A *datum feature* is a real physical part of the object which may have surface variations.

DATUM PLANE

A *datum plane* is an imaginary plane which contacts the datum feature at the highest points of variation, Figure 7-1. One or more datum planes may be used to establish positional relationships on a part. These planes are identified as primary, secondary, or tertiary datums.

Primary datum planes are developed by establishing three points of contact on the primary datum surface. The contact points must not be in the same line, Figure 7-2.

Secondary datum planes are established on the secondary datum feature. The secondary plane is perpendicular to the primary plane. Two points of contact are used to establish the secondary datum plane, Figure 7-3.

Tertiary datum planes are perpendicular to both the primary and secondary datum planes. One point of contact is used to establish the tertiary datum plane, Figure 7-4.

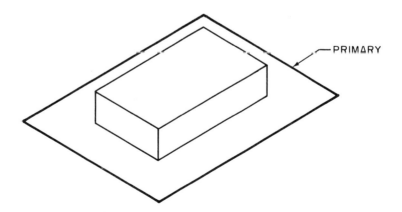

Fig. 7-2 Primary datum plane

60 Unit 7 Geometric Tolerances — Datums

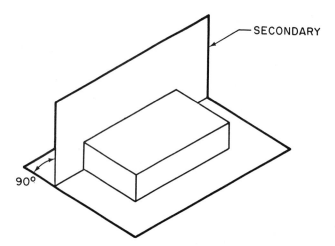

Fig. 7-3 Secondary datum plane

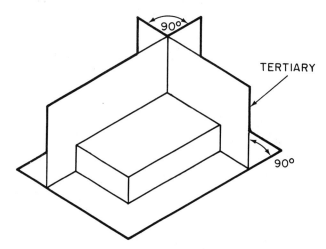

Fig. 7-4 Tertiary datum plane

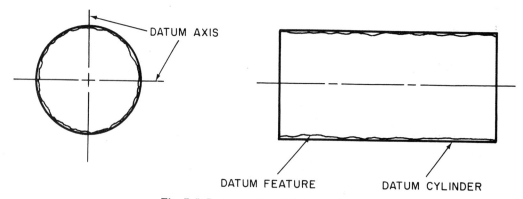

Fig. 7-5 Datum axis and datum cylinder

DATUM CYLINDER

A *datum cylinder* is a theoretically exact form profile. The datum is formed by contact with the high points of the datum feature. A datum cylinder may be internal or external as in a hole or cylindrical shaft, Figure 7-5.

DATUM AXIS

A *datum axis* is the theoretical exact center line of a datum cylinder, Figure 7-5.

DATUM IDENTIFICATION SYMBOL

Datums must be identified on drawings with a datum symbol. These symbols indicate the datum surface being referenced. The symbol used is a capital letter enclosed in a box (frame). The letter has a dash on each side of it to indicate that it applies to a datum feature, Figure 7-6.

Fig. 7-6 Datum identification symbol

If datum symbols are located on extension lines, the datum applies to that feature only. If the symbol appears on a dimension line, it applies to the entire dimension, Figure 7-7.

FEATURE CONTROL SYMBOLS

Feature control symbols are used in geometric dimensioning to specify tolerances applied to a part feature. The symbols eliminate the need for written notes on drawings. Tolerances

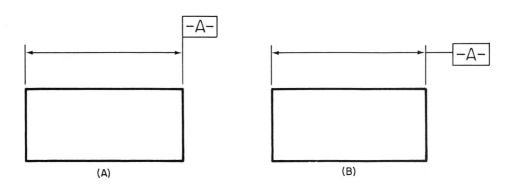

Fig. 7-7 Placement of datum symbols

TOLERANCE	SYMBOL	CHARACTERISTIC
FORM FOR INDIVIDUAL FEATURES	▱	FLATNESS
	—	STRAIGHTNESS
	○	ROUNDNESS OR CIRCULARITY
	⌭	CYLINDRICITY
PROFILE FOR INDIVIDUAL OR RELATED FEATURES	⌒	PROFILE OF A LINE
	⌓	PROFILE OF A SURFACE
ORIENTATION	∠	ANGULARITY
	⊥	PERPENDICULARITY
	∥	PARALLELISM
LOCATION	⌖	POSITION
	◎	CONCENTRICITY
RUNOUT	↗	CIRCULAR RUNOUT
	↗↗	TOTAL RUNOUT
MODIFIERS	Ⓜ	MAXIMUM MATERIAL CONDITION (MMC)
	Ⓢ	REGARDLESS OF FEATURE SIZE (RFS)
	Ⓛ	LEAST MATERIAL CONDITION (LMC)

Fig. 7-8 Chart of feature control symbols

Unit 7 Geometric Tolerances — Datums 63

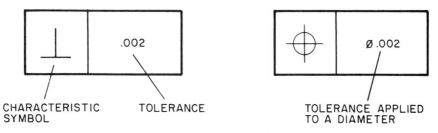

Fig. 7-9 Typical feature control symbol

specified may be tolerances of position or form. The symbols used and the characteristics they control are shown in Figure 7-8.

The method of applying feature control symbols and tolerances to drawings is the same as that used for datums. Feature control symbols are enclosed in a frame. The frame may be divided into two or more separate parts. The first space within the frame shows the geometric symbol. The second space specifies the tolerance applied to the feature. If the tolerance applies to a diameter, the symbol for diameter precedes the tolerance dimension. Figure 7-9 shows a typical feature control symbol.

A leader line is used to connect the control feature frame with the part feature being controlled, Figure 7-10.

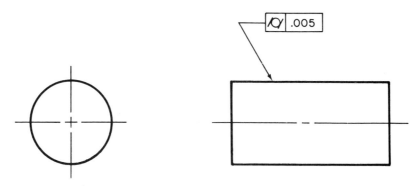

Fig. 7-10 Use of leader line with control feature frame

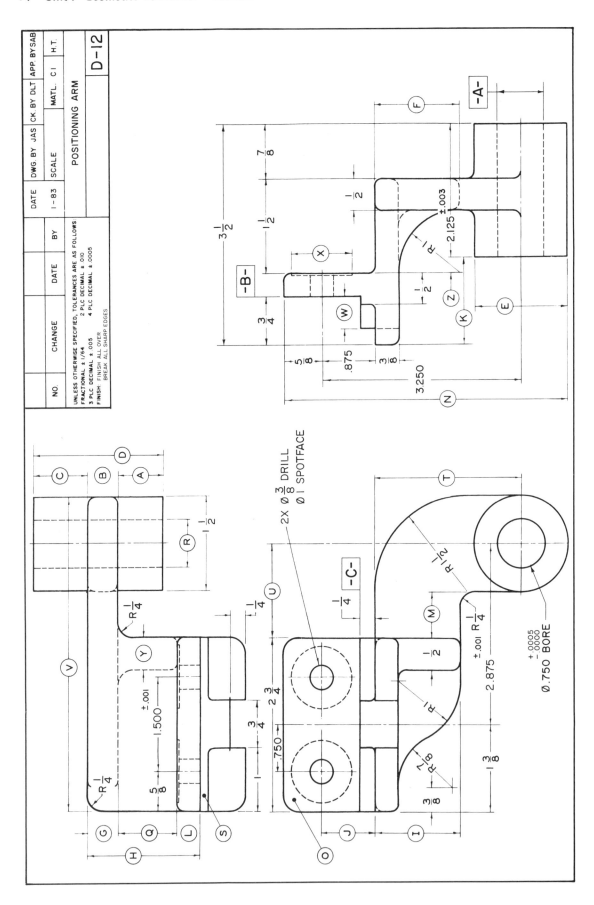

ASSIGNMENT D-12: POSITIONING ARM

1. What surface in the top view is used to establish the tertiary datum? _____

2. What diameter in the top view is used to establish the primary datum? _____

3. What surface in the front view is used to establish the secondary datum? _____

4. Projected surfaces may be identified by determining distances. Determine the distances indicated by each of the letters to the right.

5. What is the nominal size of the bored hole? _____

6. What is the maximum material condition (MMC) of the two drilled holes? _____

7. What surface in the front view is the "specified datum" for datum B? _____

Ⓐ = _____
Ⓑ = _____
Ⓒ = _____
Ⓓ = _____
Ⓔ = _____
Ⓕ = _____
Ⓖ = _____
Ⓗ = _____
Ⓘ = _____
Ⓙ = _____
Ⓚ = _____
Ⓛ = _____

Ⓜ = _____
Ⓝ = _____
Ⓞ = _____
Ⓡ = _____
Ⓣ = _____
Ⓤ = _____
Ⓥ = _____
Ⓦ = _____
Ⓧ = _____
Ⓨ = _____
Ⓩ = _____

unit 8

GEOMETRIC TOLERANCES — LOCATION AND FORM

In the previous unit the basic symbols and terminology used in geometric dimensioning were discussed. This unit provides a greater understanding of each characteristic and how tolerances are identified.

MODIFIERS

Maximum Material Condition (MMC) Ⓜ

The *maximum material condition* of a part exists when a feature contains the maximum material allowed. An example would be a pin or shaft at its high limit dimension or a slot or hole at its lowest limit, Figure 8-1. Maximum material condition is specified by the modifier symbol Ⓜ. It is also abbreviated with the letters MMC.

The maximum material condition applies when:

1. Two or more features are interrelated with respect to location or form. For example, a hole and an edge or two holes, etc. At least one of the related features is to be one of size.
2. The feature to which the MMC applies must be a feature of size. For example, a hole, slot, or pin with an axis.

Regardless of Feature Size (RFS) Ⓢ

The regardless of feature size symbol is no longer required on industrial drawings. However it may still be found on various prints or may be used by some companies. RFS is a condition where the tolerance of form or position must be met regardless of where the feature lies within the size tolerance. The modifier symbol for RFS is Ⓢ.

When modifiers are specified on a drawing they appear in the same box and to the right of the tolerance, Figure 8-2.

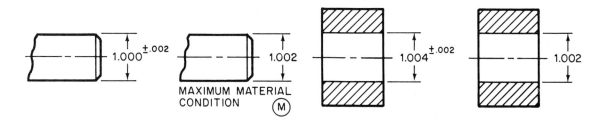

Fig. 8-1 An example of maximum material condition

68 Unit 8 Geometric Tolerances – Location and Form

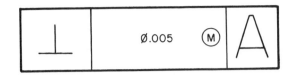

Fig. 8-2 Specifying modifiers on a drawing

LOCATION TOLERANCES

Position

True position is the term applied to the exact or perfect location of a feature. The true position is located with reference to one or more datums. The position tolerance is the maximum amount of variation allowed. From true position, the amount of permissible tolerance is called a *tolerance zone*. For cylindrical features this zone is a diameter within which the axis of the feature must lie, Figure 8-3.

Concentricity

Concentricity refers to the condition of two features sharing a common axis. An example would be a stepped shaft where two diameters share a common centerline, Figure 8-3. The concentricity tolerance is the diameter of the concentricity tolerance zone within which the feature axis must lie.

FORM TOLERANCES

Form tolerances specify the allowable deviation from perfect form as specified on the print, Figure 8-4. These form tolerances are specified where part features are critical to interchangeability.

Flatness

Flatness is the condition of having all elements of a surface in one plane, Figure 8-4A. The tolerance for flatness is defined as the dimensional area formed by two flat planes. The entire surface of the feature including variations must fall in this zone.

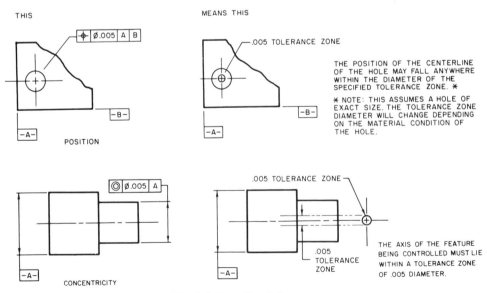

Fig. 8-3 Location tolerances

Straightness ———

Straightness is different than flatness and should not be confused. *Straightness* refers to an element of a surface being in a straight line, Figure 8-4B. The tolerance for straightness specifies a zone of uniform width along the length of the feature. All points of the feature as measured along that line must fall within that zone.

Roundness ◯

Roundness is the condition where all points on the feature surface are an equal distance from the center, Figure 8-4C. The tolerance zone is formed by two concentric circles. The actual surface elements must lie between these two circles at any place of cross section.

Cylindricity ⌭

Cylindricity is the condition where all elements of a surface of revolution form a cylinder. The tolerance zone is defined by two concentric cylinders along the length of the feature, Figure 8-4D. All points of the part feature must fall between these cylinders.

Profile of a Line ⌒

The *profile of a line* is the feature profile as measured along a line, Figure 8-4E. The tolerance of the profile is the variation from perfect form. All points along the part feature must fall between parallel lines of the perfect profile.

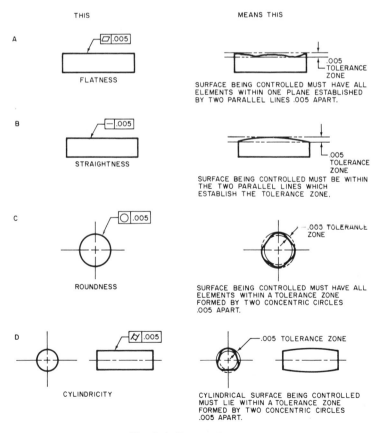

Fig. 8-4 Form tolerance

Unit 8 Geometric Tolerances — Location and Form

THIS   MEANS THIS

E   ⌒ .005

PROFILE OF A LINE

.005 TOLERANCE ZONE

THE LINE ELEMENT AT ANY CROSS SECTION MUST LIE WITHIN THE BOUNDARIES FORMED BY TWO PARALLEL LINES OF TRUE PROFILE .005 APART.

F   ⌓ .005

PROFILE OF A SURFACE

.005 TOLERANCE ZONE

SURFACE PROFILE BEING CONTROLLED MUST HAVE ALL ELEMENTS WITHIN A TOLERANCE ZONE FORMED BY TWO PARALLEL PLANES OF TRUE PROFILE .005 APART.

G   ∥ .005 A

PARALLELISM
-A-

.005 TOLERANCE ZONE
-A-

SURFACE BEING CONTROLLED MUST LIE WITHIN A TOLERANCE ZONE FORMED BY TWO PLANES .005 APART AND PARALLEL TO DATUM A.

H   ⊥ .005 A

PERPENDICULARITY
-A-

.005 TOLERANCE ZONE
-A-

SURFACE BEING CONTROLLED MUST LIE WITHIN A TOLERANCE ZONE FORMED BY TWO PARALLEL LINES .005 APART AND PERPENDICULAR TO DATUM A.

THIS   MEANS THIS

I   ∠ .005 A   15°

ANGULARITY
-A-

.005 TOLERANCE ZONE
-A-

THE SURFACE BEING CONTROLLED MUST LIE WITHIN A TOLERANCE ZONE FORMED BY TWO PARALLEL PLANES .005 APART AND INCLINED AT 15° TO DATUM A.

J   ↗ .005 A

CIRCULAR RUNOUT
-A-

-A-

AT ANY POINT ON THE SURFACE THE TOTAL INDICATOR READING (TIR) MUST LIE WITHIN .005 WHEN THE PART IS ROTATED 360°.

K   ↗↗ .005 A

TOTAL RUNOUT
-A-

-A-

THE ENTIRE SURFACE BEING CONTROLLED MUST LIE WITHIN .005 TOTAL INDICATOR READING (TIR) WHEN THE PART IS ROTATED 360°.

Fig. 8-4 (continued)

Profile of a Surface ⌒

The *profile of a surface* is much like the profile of a line. However, the definition is broadened to cover the entire feature surface, Figure 8-4F. All points of the feature surface must fall within the tolerance zone of perfect profile.

Parallelism //

Parallelism refers to a surface, line, or axis which is an equal distance from a datum plane or axis at all points, Figure 8-4G. The tolerance zone specified is defined by two planes parallel to a datum plane. It may also be defined as a cylindrical tolerance zone parallel to a datum axis.

Perpendicularity ⊥

Perpendicularity is the condition where a feature is 90 degrees from a datum plane or axis, Figure 8-4H. All points of the feature surface must fall within the zone formed by lines perpendicular to the datum.

Angularity ∠

Angularity is the condition of a surface, axis, or center plane which is at an angle other than 90 degrees from a center plane or axis, Figure 8-4I. The tolerance zone is formed by two parallel lines inclined at the exact angle specified.

Runout

Circular Runout (↗) is the deviation of the part feature at any measuring position when rotated on a datum axis, Figure 8-4J. A dial indicator is used to read runout on a feature. The tolerance zone is formed by two parallel lines within which the total feature runout must fall.

Total Runout (↗↗) is the deviation of the entire surface at any measuring point within the specified tolerance zone.

REVIEW OF SYMBOLOGY

To the following questions, add the necessary datums (example: |-A-|) and feature control symbols (example: | ⌀ | .005 | A |) to make the statement correct.

1. Make the top flat to within .005.

2. Make the end perpendicular to the bottom within .005 maximum material condition.

3. Make the periphery of this cylinder round to within .005.

4. Make the top and bottom surfaces flat and equal distance from each other within .005.

5. Make the 30-degree angle correct to within .005 of the bottom surface.

6. Make this shaft concentric to the centerline within .005.

7. Make this shaft cylindrical within .005 to the bore.

8. Make the two top surfaces parallel to the bottom within .002.

9. Make the profile of the top surface within .004 of the end.

10. Make each diameter of the shaft be within .002 for runout with the centerline.

Unit 8 Geometric Tolerances — Location and Form 73

11. Make the top parallel, the right end perpendicular, and left angular to the bottom datum within .003.

12. Make the shaft straight within .003 MMC.

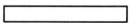

13. Make the top parallel to the bottom within .006.

14. Make runout on the bore, faces, and diameters within .002 of the centerline.

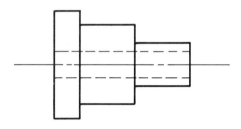

15. Make true position of the mating parts to within .002 at MMC. The pins are pressed into the block and the holes are reamed.

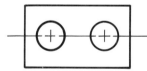 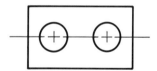

16. Show the difference of the tolerances that true position gives a part when added to the following diagram:

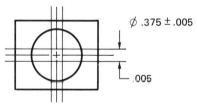

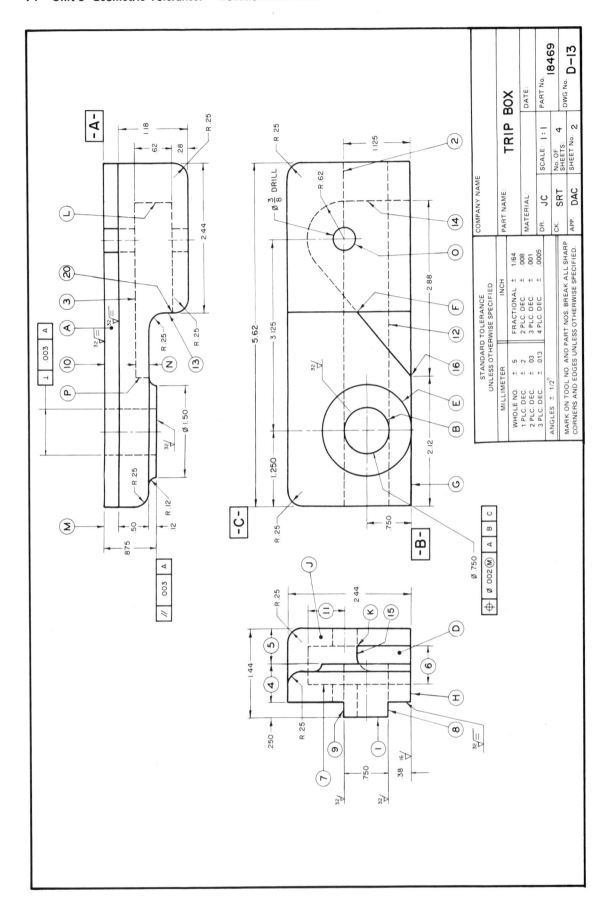

ASSIGNMENT D-13: TRIP BOX

1. What line in the top view represents surface ①?
2. Locate surface Ⓐ in the left and front views.
3. Locate surface ⑧ in the front view.
4. How many surfaces are to be finished?
5. What line in the left view represents surface ③?
6. What is the center distance between holes Ⓑ and Ⓞ?
7. Determine distance ④.
8. Determine distance ⑤.
9. Determine distance ⑥.
10. Determine distance ⑪.
11. Locate surface Ⓙ in the top view.
12. What surface of the left view does line ⑭ represent?
13. What point in the front view represents line ⑮?
14. What is the thickness of boss Ⓔ?
15. Locate surface Ⓖ in the left view.
16. Locate point Ⓚ in the top view.
17. Locate surface Ⓓ in the top view.
18. Determine distance Ⓜ.
19. Determine distance Ⓝ.
20. What point or line in the top view does point ⑯ represent?
21. What tolerance of parallelism is allowed on the surface of boss Ⓔ?
22. What surface in the left view is used as the primary datum?
23. What does the geometric symbol Ⓜ indicate?
24. What surface finish is required on the surface shown by hidden line ②?
25. The .750 hole must be perpendicular within .003 with respect to which datum?

unit 9

SCREW THREADS

Screw threads play an important part in industry. They are used for fastening parts together, for making adjustments, and for transmitting power. Standard screw threads are available in various sizes and forms. The form refers to the shape of the thread.

SCREW THREAD FORMS

The *form* of a screw thread is the profile or side view of the thread. A number of screw thread forms are used for industrial applications. The form selected often depends on what the screw thread is to be used for. A thread form used as a fastener, such as a bolt, usually differs from one used to transmit power. Figure 9-1 shows the most common standard thread forms.

The most frequently used thread form is Unified standard. The Unified form is used in the United States, Canada, and Great Britain. The Unified thread has all but replaced the American National form because it is easier to produce. The only difference in the form is the shape of the crest and root. The various thread forms and their applications are:

Sharp V. The sharp V thread is seldom used except for some special purposes. The sharp pointed crest and root make this thread susceptible to damage. Sharp V threads are often used on brass pipe.

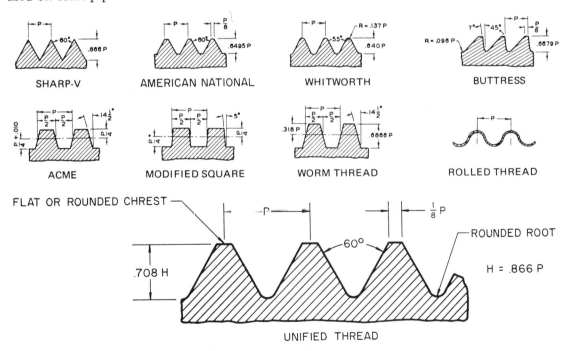

Fig. 9-1 Standard thread forms

American National. The American National screw thread has a flat crest and a flat root. It is stronger than a sharp V and less susceptible to damage. The American National form has been used most often on fasteners.

Unified Form. The Unified form of thread was developed to replace the American National form. The crest of the Unified form may be flat or rounded. The root of the thread is rounded. Otherwise, Unified and American National thread forms appear very similar.

Acme Form. Acme thread forms are very strong and are used to transmit power. A large flat at the crest and root are characteristics of Acme threads. Typical applications for Acme forms are lathe lead screws and vise screws.

Worm Form. Worm threads are similar in shape to Acme threads. They are used to transmit power and motion to worm wheels.

Buttress Form. Buttress threads are used to transmit power in one direction. They are capable of handling high stress as in screw jack applications.

Whitworth Form. Whitworth threads formerly were the British standard threads. The Unified form has replaced the Whitworth in most applications.

Rolled Form. Rolled threads or knuckle threads are formed from sheet metal or cast and are used for electrical parts and screw parts. The screw shells of electric light bulbs, lamp bases, and bottle tops are examples.

SCREW THREAD TERMINOLOGY

The definitions of terms relating to screw threads are illustrated in Figures 9-2 and 9-3.

Screw thread — A ridge of uniform section in the form of a helix on the external or internal surface of a cylinder.

External thread — A thread on the outside of a member. Example: a bolt.

Internal thread — A thread on the inside of a member. Example: a thread in a nut.

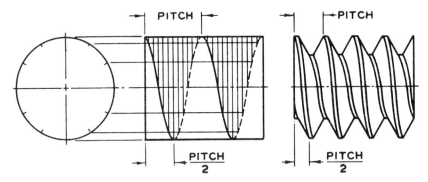

Fig. 9-2 A helix and a screw thread

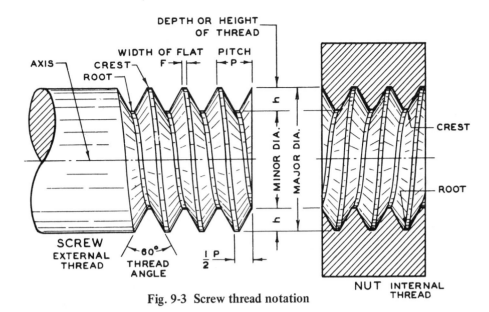

Fig. 9-3 Screw thread notation

Major diameter — The largest diameter of the thread of a screw or nut.

Minor diameter — The smallest diameter of the thread of a screw or nut.

Pitch — The distance from a point on a screw thread to a corresponding point on the next thread measured parallel to the axis.

Lead — The distance a screw thread advances axially in one turn. On a single-thread screw the lead and the pitch are the same; on a double-thread screw, the lead is twice the pitch; on a triple-thread screw, the lead is three times the pitch.

Angle of thread — The angle included between the sides of the thread measured in an axial plane.

Crest — The surface of a thread joining two sides at the major diameter of a screw and the minor diameter of a nut.

Root — The surface of a thread joining two sides at the minor diameter of a screw and the major diameter of a nut.

Depth — The distance from the root of a thread to a line between two adjacent crests measured perpendicular to the axis.

Axis — The centerline running lengthwise through a screw.

Screw threads required on a workpiece must be specified on the print. The specifications should include:

- The outside diameter or major diameter of the thread
- The thread series
- The number of threads per inch of thread
- The class of fit

Thread specifications are called out using a leader and note. The leader arrow points to the thread, and the note contains the specifications. These specifications are presented in a sequence of information described later in the unit.

UNIFIED NATIONAL THREAD SERIES

The Unified National threads have been classified into six standard thread series. They are American National coarse, fine, extra-fine, 8-pitch, 12-pitch, and 16-pitch thread series.

The Unified National coarse thread series (UNC) is recommended for general use in machine construction. Sizes range from No. 1 (.073-inch diameter) to 4-inch diameter.

The fine-thread series (UNF) is recommended where conditions require a fine thread. Sizes range from No. 0 (.060-inch diameter) to 1 1/2-inch diameter.

The extra-fine series (UNEF) is recommended where thinwalled material is to be threaded and where depth of thread must be held to a minimum. Sizes range from 1/4-inch diameter to 2-inch diameter.

The 8-pitch thread series (8UN) has eight threads per inch for all diameters. It is generally used on bolts for high-pressure pipe flanges, cylinder head studs, and similar fastenings. Sizes range from 1-inch diameter to 6-inch diameter.

The 12-pitch thread series (12UN) has 12 threads per inch for all diameters. It is used in boiler practice and for thin nuts and threaded collars. Sizes range from 1/2-inch diameter to 6-inch diameter.

The 16-pitch thread series (16UN) has 16 threads per inch for all diameters. It is used on threaded adjusting collars. Sizes range from 3/4-inch diameter to 4-inch diameter.

The thread diameters and the number of threads per inch for the six standard thread series are given in Figure 9-4.

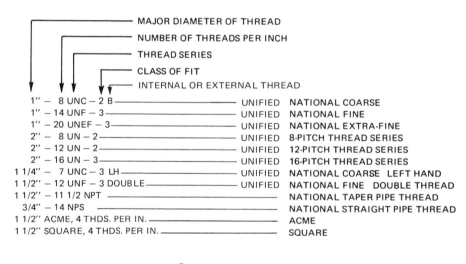

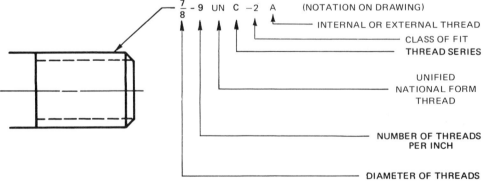

Fig. 9-4 Thread specifications

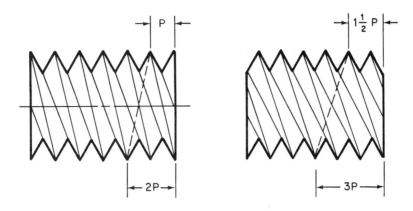

Fig. 9-5 Multiple threads

MULTIPLE THREADS

A *screw thread* as defined is a uniform ridge cut on the diameter of a cylinder. Threads may be single or multiple depending on the requirements of the thread. *Single threads* have one ridge. *Multiple threads* have two or more ridges which run side by side.

Multiple threads are normally double or triple threads. A *double thread* has a lead which is twice the pitch. A *triple thread* has a lead which is 1 1/2 times the pitch, Figure 9-5.

Multiple threads are used where a screw or mating part must be rapidly advanced along the thread. One turn of a double thread will advance it twice as far as one turn of a single thread. A triple thread will advance three times as far as a single thread per turn. Multiple threads are often found on toothpaste caps, valve stems, or other such applications. These types of threads are used for motion rather than power.

CLASSIFICATION OF FITS

Five distinct classes of screw thread fits have been established. This is to insure interchangeable manufacture of screw thread parts throughout the country.

Class 1. Recommended only for screw thread work where clearance between mating parts is essential for rapid assembly and where shake or play is allowed.

Class 2. Represents a high quality of commercial screw thread product. It is recommended for the great bulk of interchangeable screw thread work.

Class 3. Represents an exceptionally high grade of commercially threaded product. It is recommended only in cases where the high cost of precision tools and continual checking of tools and product is warranted.

Class 4. Intended to meet very unusual requirements more exacting than those for which class 3 is intended. It is a selective fit if initial assembly by hand is required. It is not, as yet, adaptable to quantity production.

Class 5. Includes interchangeable screw thread work consisting of steel studs set in hard materials (cast iron, steel, bronze, etc.) where a wrench-tight fit is required.

SYMBOLS FOR IDENTIFYING THREAD SPECIFICATIONS

A system of letters and numbers are used to identify thread specifications on a print. The sequence used states:

1. The major or outside diameter of the thread
2. The number of threads per inch
3. The thread series
4. The class of fit

Thread identification symbols are shown in Figure 9-4.

Threads may be cut as either a right-hand thread or a left-hand thread. When a right-hand thread is specified, no special symbol is put on the drawing. When a left-hand thread is specified, the symbol *LH* is placed after the designation of the thread.

Internal and external threads are represented on drawings in one of three ways. They may be drawn pictorially in schematic or using a simplified form.

PICTORIAL REPRESENTATION

Pictorial representations show the thread form very close to how it actually appears, Figure 9-6. The detailed shape is very pleasing to the eye. However, the task of drawing pictorials is very difficult and time consuming. Therefore, pictorial representations are rarely used on threads of less than one inch in diameter.

SCHEMATIC REPRESENTATION

Schematic thread representation does not show the outline of the thread shape, Figure 9-7. Instead two parallel lines are drawn at the major diameter. The crest and root lines are drawn at right angles to the thread axis instead of sloping. The root lines are drawn thicker than the crest lines. In actual drafting practice, the crest and root lines are spaced by eye to the approximate pitch, and they may be of equal width if preferred. The schematic thread symbol for threads in section does have the 60-degree thread outline. On one edge the thread outline is advanced one-half of the pitch.

The ends of the screw threads are chamfered at 45 degrees to the thread depth. This protects the starting thread and permits the engaging nut to start easily.

Internal schematic thread representation does not show the vertical crest and root lines. Instead two parallel hidden lines are used to represent these points. However, vertical crest and root lines are used when a section view is shown, Figure 9-8.

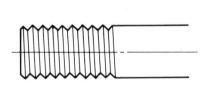

Fig. 9-6 Pictorial thread representation

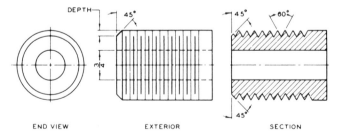

Fig. 9-7 External schematic thread representation

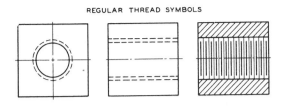

Fig. 9-8 Internal schematic thread representation

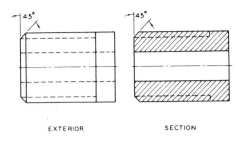

Fig. 9-9 Simplified external thread symbols

SIMPLIFIED THREAD REPRESENTATION

Simplified thread symbols are used to further reduce drafting time. The thread outline and the crest and root lines are not drawn. Two dotted lines parallel to the axis are drawn to indicate the depth and the length of the threads, Figure 9-9.

Internal simplified threads are represented the same as the schematic representations, Figure 9-10. However, in sectional views the crest and root lines are omitted. Instead, lines are drawn parallel to the thread axis to represent the major and minor diameters. The lines representing the major diameter are made with short dashes.

REPRESENTING TAPPED HOLES

Tapped holes may be represented in any one of the three forms previously mentioned. The most frequently used are the schematic or simplified.

Tapped or threaded holes may go all the way through a piece or only part way. Holes which are tapped through are represented as shown in Figure 9-11.

Fig. 9-10 Simplified internal thread symbols

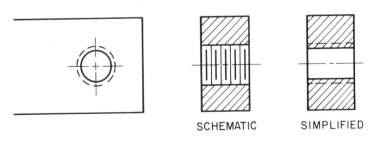

Fig. 9-11 Hole which is tapped through

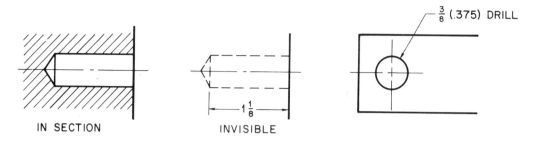

Fig. 9-12 Representing tap drill holes

Threads which are not tapped through are shown in combination with the tap drill hole. Tap drill holes are represented with hidden lines showing the outside diameter of the drill. This same surface is used to represent the major diameter of the screw thread. The bottom of the tap drill hole is pointed to represent the drill point, Figure 9-12.

Tapped holes may be shown threaded to the bottom of the drill hole or to a specified depth, Figure 9-13. The specified depth is called out on the print, Figure 9-14.

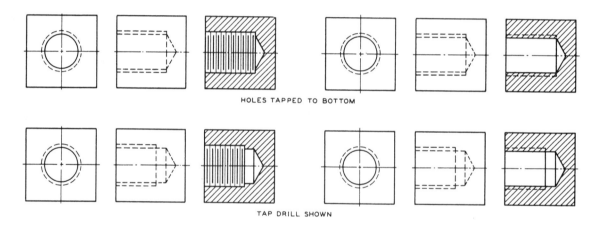

Fig. 9-13 Internal thread symbols

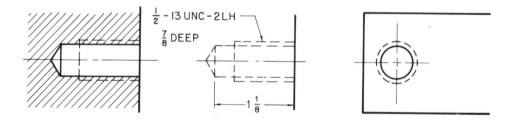

Fig. 9-14 Simplified representation of tapped holes

ASSIGNMENT D-14: CROSS HEAD

Make a freehand sketch showing a top view of the Cross Head. Place the sketch in the space provided below.

NOTE: Several designs are possible for the sketch of the top view.

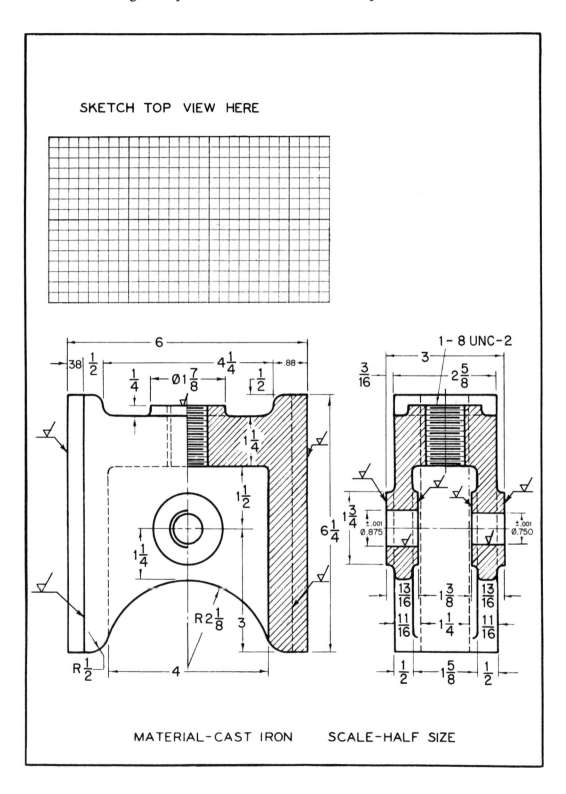

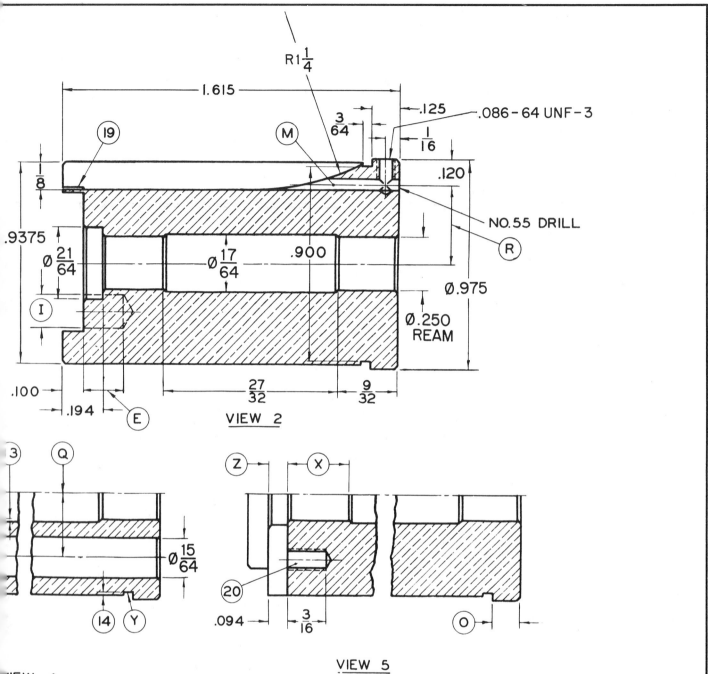

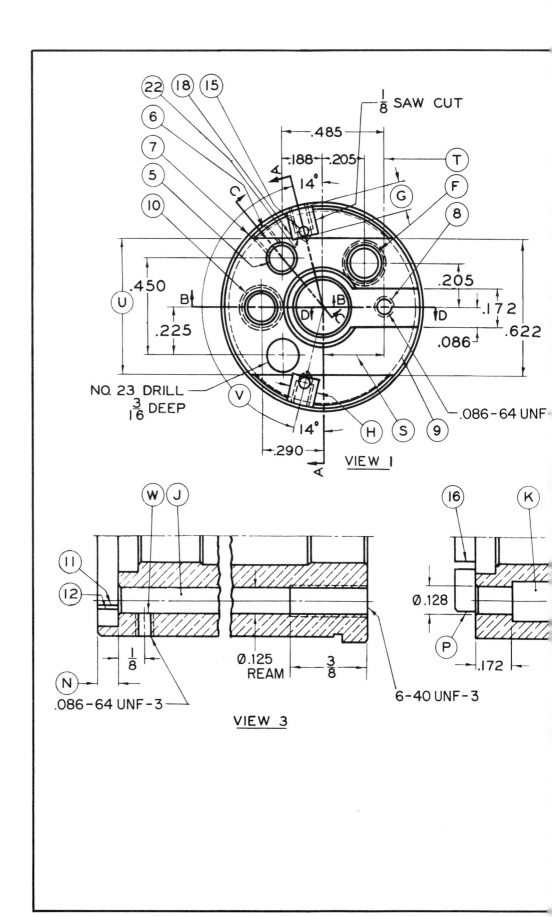

ASSIGNMENT D-15: SPINDLE BEARING

NOTE: Section views of the interior of an object show more clearly the details which otherwise would be difficult to interpret in the regular views. The number of section views taken depends upon the complexity of the part.

In the drawing of the Spindle Bearing, four section views are needed to bring out the details of construction. When the sections are not labeled, it becomes necessary to note several characteristics of each section view, and then locate one or more of them in an ordinary view, as a help in determining where the section is taken. For example, in view 4 there are no threaded holes to identify the section as in view 3, but there are two chamfered edges, holes of two different sizes, and the lines at 16, all of which should aid in determining where the section is taken.

1. Determine distance Q. _____
2. Determine distance R. _____
3. Determine distance S. _____
4. What is distance E? _____
5. What is the diameter of the hole at F? _____
6. Determine distance G. _____
7. What would you assume distance H to be? _____
8. What is the diameter at I? _____
9. Identify hole J in view 1. _____
10. Identify hole K in view 1. _____
11. Identify hole M in view 1. _____
12. Identify hole 8 in another view. _____
13. Determine distance N. _____
14. Determine distance O. _____
15. Line P is represented by a point in view 1. Identify the point. _____
16. What cutting plane line in view 1 indicates the section shown in view 2? _____
17. View 3 is a section taken from view 1. Indicate the section from which it was taken. _____
18. View 4 is a section taken from view 1. Locate the portion of view 1 from which it was taken. _____
19. Determine distance T. _____
20. Determine distance U. _____
21. Determine angle V. _____
22. Locate hole W on view 1. _____
23. Determine distance X. _____
24. Identify line Y on view 1. _____
25. Determine distance Z. _____
26. Lines 11 and 12 are represented by points or lines in view 1. Identify the points or lines. _____
27. Point 22 is shown by a point or line in view 4. Identify the location. _____
28. Point 18 is shown by a point or line in view 2. Identify the location. _____
29. Determine the depth of the slot at 14. _____
30. Determine the depth of the recess at 13. _____
31. What class of fit is required on the threaded holes? _____
32. How many .086-64 UNF-3 holes are required? _____
33. What thread form is required? _____
34. What thread series is required? _____
35. What type of thread representation is used to show the screw threads? _____

unit 10

THREADED FASTENERS

Threaded fasteners are used for assembly, clamping, or adjusting purposes in machine work. Common threaded fasteners include a variety of screws, nuts, bolts, or studs. This unit describes the five basic threaded fasteners and their uses.

MACHINE SCREWS

Machine screws are available in a variety of thread sizes and lengths. They are used frequently where small diameter fasteners are required for general assembly work. Machine screws are much like machine bolts or cap screws in appearance. However, machine screws are generally smaller and have slotted or cross-slotted heads. The head shape selection is determined by the screw application. Figure 10-1 shows the common machine screws used.

CAP SCREWS

Cap screws are used for assembly purposes. They are stronger and more precise than machine screws. Cap screws are often used to hold two pieces together. The body of the cap screw passes through a clearance hole in one piece and threads into the mating part. Cap screws may have slotted, hex, or socket hex drive heads. Figure 10-2 shows the common cap screws used.

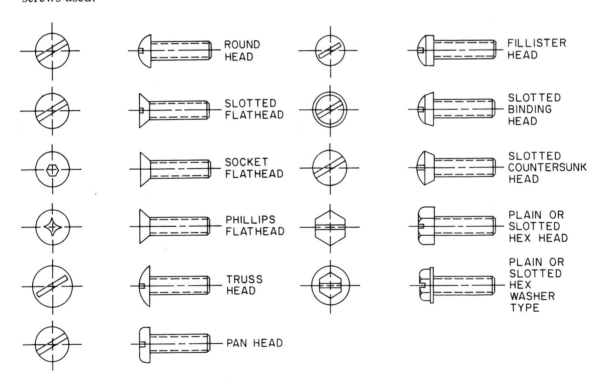

Fig. 10-1 Machine screws

88

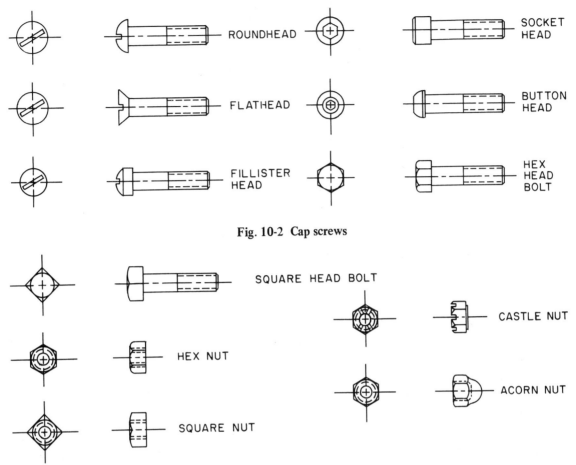

Fig. 10-2 Cap screws

Fig. 10-3 Nuts and bolts

MACHINE BOLTS

Machine bolts are used to clamp two or more parts together. They are not as precise as cap screws and are not available in as large a variety of head forms. Machine bolts have either square or hex-shaped heads.

Bolts are designed to pass through clearance holes in assembled parts. The tightening or releasing of a bolt is normally accomplished by use of a torquing nut. Standard bolt and nut forms are shown in Figure 10-3.

STUD BOLTS

Stud bolts or studs are headless bolts with threads on each end. One end is threaded into a tapped hole, while a clamping nut is used on the other end. Studs are frequently used in clamping applications such as securing a part to a milling table. They are also commonly used to secure equipment to a floor or base. Figure 10-4 shows a standard stud bolt.

Fig. 10-4 Stud bolt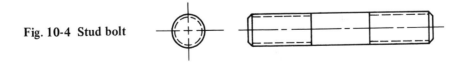

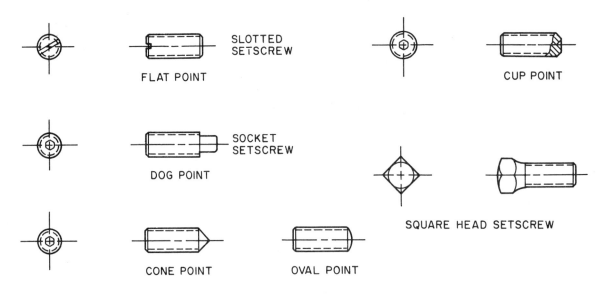

Fig. 10-5 Setscrews

SETSCREWS

Setscrews are used to prevent motion or slippage between two parts, such as pulleys or collars on a shaft. Setscrews are usually heat treated for added strength to resist wear. They may be either head or headless with a variety of point forms available. Some basic setscrews and point variations are shown in Figure 10-5.

WASHERS

Washers are accessories which are used with screws, nuts, bolts, and studs. Washers help to distribute clamping pressure over a wider area. They also prevent surface marring which may result from the tightening of the bolt head or nut. The most common types of machine washers are the plain flat washer and the spring lock washer. Spring lock washers are used to prevent *backing off* or loosening of threaded fastener assemblies. Tables 10-1 and 10-2 list the basic washer sizes and illustrate the shape of each. Figure 10-6 shows some typical fastener assemblies.

THREADED FASTENER SIZE

The size of a threaded fastener such as a screw, bolt, or stud is determined by the nominal thread diameter. The body is used to designate the dimension of length. However, the thread length may vary with the diameter or style of fastener. Figure 10-7 illustrates the common size designations of a typical fastener.

Table 10-1 Dimensions of Preferred Sizes of Type A Plain Washers** (ANSI B18.22.1-1965)

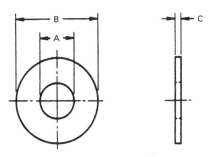

Nominal Washer Size***			Inside Diameter A			Outside Diameter B			Thickness C		
			Basic	Tolerance Plus	Tolerance Minus	Basic	Tolerance Plus	Tolerance Minus	Basic	Max	Min
—	—		0.078	0.000	0.005	0.188	0.000	0.005	0.020	0.025	0.016
—	—		0.094	0.000	0.005	0.250	0.000	0.005	0.020	0.025	0.016
—	—		0.125	0.008	0.005	0.312	0.008	0.005	0.032	0.040	0.025
No. 6	0.138		0.156	0.008	0.005	0.375	0.015	0.005	0.049	0.065	0.036
No. 8	0.164		0.188	0.008	0.005	0.438	0.015	0.005	0.049	0.065	0.036
No. 10	0.190		0.219	0.008	0.005	0.500	0.015	0.005	0.049	0.065	0.036
3/16	0.188		0.250	0.015	0.005	0.562	0.015	0.005	0.049	0.065	0.036
No. 12	0.216		0.250	0.015	0.005	0.562	0.015	0.005	0.065	0.080	0.051
1/4	0.250	N	0.281	0.015	0.005	0.625	0.015	0.005	0.065	0.080	0.051
1/4	0.250	W	0.312	0.015	0.005	0.734*	0.015	0.007	0.065	0.080	0.051
5/16	0.312	N	0.344	0.015	0.005	0.688	0.015	0.007	0.065	0.080	0.051
5/16	0.312	W	0.375	0.015	0.005	0.875	0.030	0.007	0.083	0.104	0.064
3/8	0.375	N	0.406	0.015	0.005	0.812	0.015	0.007	0.065	0.080	0.051
3/8	0.375	W	0.438	0.015	0.005	1.000	0.030	0.007	0.083	0.104	0.064
7/16	0.438	N	0.469	0.015	0.005	0.922	0.015	0.007	0.065	0.080	0.051
7/16	0.438	W	0.500	0.015	0.005	1.250	0.030	0.007	0.083	0.104	0.064
1/2	0.500	N	0.531	0.015	0.005	1.062	0.030	0.007	0.095	0.121	0.074
1/2	0.500	W	0.562	0.015	0.005	1.375	0.030	0.007	0.109	0.132	0.086
9/16	0.562	N	0.594	0.015	0.005	1.156*	0.030	0.007	0.095	0.121	0.074
9/16	0.562	W	0.625	0.015	0.005	1.469*	0.030	0.007	0.109	0.132	0.086
5/8	0.625	N	0.656	0.030	0.007	1.312	0.030	0.007	0.095	0.121	0.074
5/8	0.625	W	0.688	0.030	0.007	1.750	0.030	0.007	0.134	0.160	0.108
3/4	0.750	N	0.812	0.030	0.007	1.469	0.030	0.007	0.134	0.160	0.108
3/4	0.750	W	0.812	0.030	0.007	2.000	0.030	0.007	0.148	0.177	0.122
7/8	0.875	N	0.938	0.030	0.007	1.750	0.030	0.007	0.134	0.160	0.108
7/8	0.875	W	0.938	0.030	0.007	2.250	0.030	0.007	0.165	0.192	0.136
1	1.000	N	1.062	0.030	0.007	2.000	0.030	0.007	0.134	0.160	0.108
1	1.000	W	1.062	0.030	0.007	2.500	0.030	0.007	0.165	0.192	0.136
1-1/8	1.125	N	1.250	0.030	0.007	2.250	0.030	0.007	0.134	0.160	0.108
1-1/8	1.125	W	1.250	0.030	0.007	2.750	0.030	0.007	0.165	0.192	0.136
1-1/4	1.250	N	1.375	0.030	0.007	2.500	0.030	0.007	0.165	0.192	0.136
1-1/4	1.250	W	1.375	0.030	0.007	3.000	0.030	0.007	0.165	0.192	0.136
1-3/8	1.375	N	1.500	0.030	0.007	2.750	0.030	0.007	0.165	0.192	0.136
1-3/8	1.375	W	1.500	0.045	0.010	3.250	0.045	0.010	0.180	0.213	0.153
1-1/2	1.500	N	1.625	0.030	0.007	3.000	0.030	0.007	0.165	0.192	0.136
1-1/2	1.500	W	1.625	0.045	0.010	3.500	0.045	0.010	0.180	0.213	0.153
1-5/8	1.625		1.750	0.045	0.010	3.750	0.045	0.010	0.180	0.213	0.153
1-3/4	1.750		1.875	0.045	0.010	4.000	0.045	0.010	0.180	0.213	0.153
1-7/8	1.875		2.000	0.045	0.010	4.250	0.045	0.010	0.180	0.213	0.153
2	2.000		2.125	0.045	0.010	4.500	0.045	0.010	0.180	0.213	0.153
2-1/4	2.250		2.375	0.045	0.010	4.750	0.045	0.010	0.220	0.248	0.193
2-1/2	2.500		2.625	0.045	0.010	5.000	0.045	0.010	0.238	0.280	0.210
2-3/4	2.750		2.875	0.065	0.010	5.250	0.065	0.010	0.259	0.310	0.228
3	3.000		3.125	0.065	0.010	5.500	0.065	0.010	0.284	0.327	0.249

*The 0.734 in., 1.156 in., and 1.469 in. outside diameters avoid washers which could be used in coin operated devices.
**Preferred sizes are for the most part from series previously designated "Standard Plate" and "SAE." Where common sizes existed in the two series, the SAE size is designated "N" (narrow) and the Standard Plate "W" (wide). These sizes as well as all other sizes of Type A Plain Washers are to be ordered by ID, OD, and thickness dimensions.
***Nominal washer sizes are intended for use with comparable nominal screw or bolt sizes.

Courtesy of the American Society of Mechanical Engineers; ANSI B18.22.1-1965 (R1975), Table 1A

Table 10-2 Dimensions of Regular Helical Spring Lock Washers[1] (ANSI B18.21.1-1972)

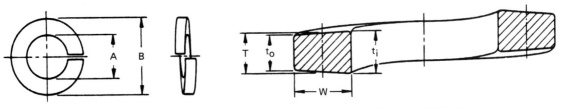

Nominal Washer Size		A Inside Diameter		B Outside Diameter	T Mean Section Thickness $\left(\dfrac{t_i + t_o}{2}\right)$	W Section Width
		Max	Min	Max[2]	Min	Min
No. 2	0.086	0.094	0.088	0.172	0.020	0.035
No. 3	0.099	0.107	0.101	0.195	0.025	0.040
No. 4	0.112	0.120	0.114	0.209	0.025	0.040
No. 5	0.125	0.133	0.127	0.236	0.031	0.047
No. 6	0.138	0.148	0.141	0.250	0.031	0.047
No. 8	0.164	0.174	0.167	0.293	0.040	0.055
No. 10	0.190	0.200	0.193	0.334	0.047	0.062
No. 12	0.216	0.227	0.220	0.377	0.056	0.070
1/4	0.250	0.262	0.254	0.489	0.062	0.109
5/16	0.312	0.326	0.317	0.586	0.078	0.125
3/8	0.375	0.390	0.380	0.683	0.094	0.141
7/16	0.438	0.455	0.443	0.779	0.109	0.156
1/2	0.500	0.518	0.506	0.873	0.125	0.171
9/16	0.562	0.582	0.570	0.971	0.141	0.188
5/8	0.625	0.650	0.635	1.079	0.156	0.203
11/16	0.688	0.713	0.698	1.176	0.172	0.219
3/4	0.750	0.775	0.760	1.271	0.188	0.234
13/16	0.812	0.843	0.824	1.367	0.203	0.250
7/8	0.875	0.905	0.887	1.464	0.219	0.266
15/16	0.938	0.970	0.950	1.560	0.234	0.281
1	1.000	1.042	1.017	1.661	0.250	0.297
1-1/16	1.062	1.107	1.080	1.756	0.266	0.312
1-1/8	1.125	1.172	1.144	1.853	0.281	0.328
1-3/16	1.188	1.237	1.208	1.950	0.297	0.344
1-1/4	1.250	1.302	1.271	2.045	0.312	0.359
1-5/16	1.312	1.366	1.334	2.141	0.328	0.375
1-3/8	1.375	1.432	1.398	2.239	0.344	0.391
1-7/16	1.438	1.497	1.462	2.334	0.359	0.406
1-1/2	1.500	1.561	1.525	2.430	0.375	0.422

[1] Formerly designated Medium Helical Spring Lock Washers.
[2] The maximum outside diameters specified allow for the commercial tolerances on cold drawn wire.
Courtesy of the American Society of Mechanical Engineers; ANSI B18.21.1-1972, Table 2

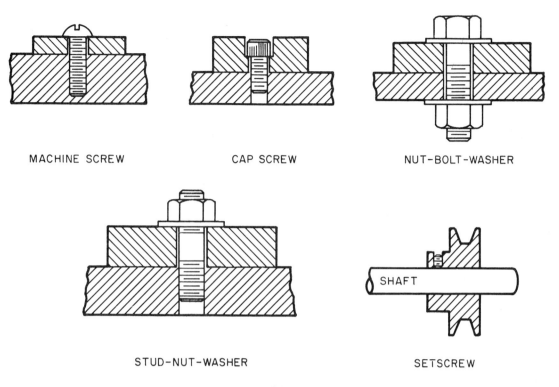

Fig. 10-6 Typical fastener assemblies

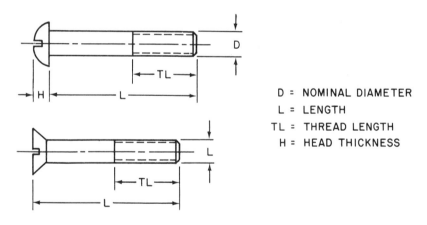

Fig. 10-7 Fastener size specifications

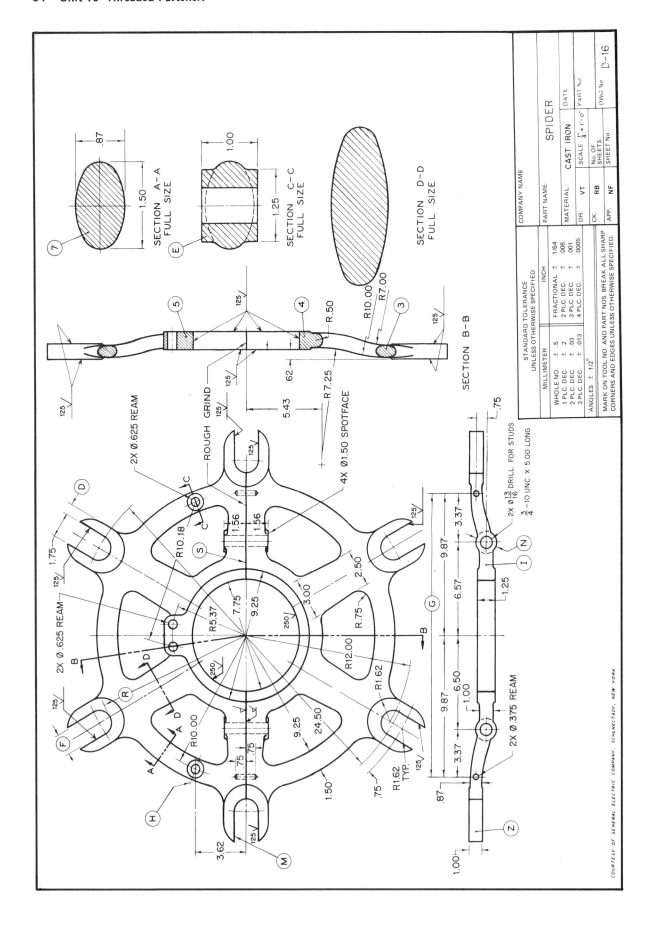

ASSIGNMENT D-16: SPIDER

1. Locate surface Ⓘ in the top view. _____

2. Locate surface Ⓩ in the top view. _____

3. What is the approximate extreme outside diameter of the Spider? _____

4. What length and diameter of studs would be used to hold the two parts of the Spider together? _____

5. Using the standard method, designate the thread size of the studs. _____

6. What type of section view is shown at B-B? _____

7. What surface finish is required on the machined surfaces? _____

8. What will be the rough dimensions of the casting at Ⓕ assuming that 1/8" overall has been allowed for finishing? _____

9. Give the finished dimension for Ⓓ. _____

10. What is the center-to-center distance Ⓖ? _____

11. What is distance Ⓡ? _____

12. What is diameter Ⓝ? _____

13. One of the rules of cross sectioning states: "No invisible edges are to be shown in cross sections." Why are the invisible lines shown in section C-C? _____

14. Where might section C-C have been taken other than on line C-C and appear the same? _____

15. How many *other* parts of the Spider have a shape whose cross section would be the same as section A-A? _____

Unit 10 Threaded Fasteners

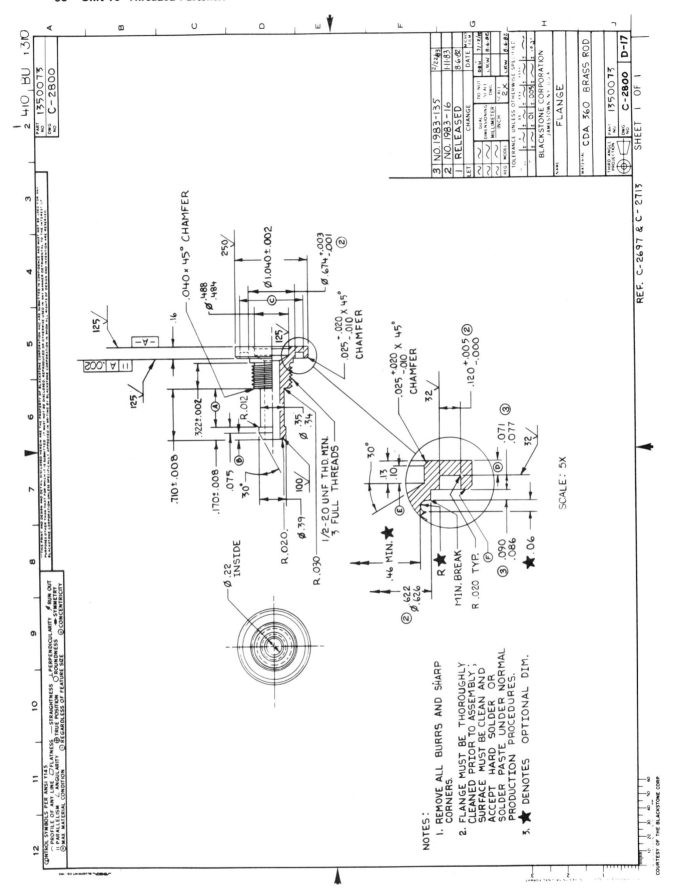

ASSIGNMENT D-17: FLANGE

1. What material is the Flange made from? _____

2. What size thread is required on the Flange? _____

3. What is the minimum thread length allowed? _____

4. What is the surface finish requirement for datum A? _____

5. What is the parallelism requirement for surface Ⓕ ? _____

6. What is the depth of the .120 wide groove in the head of the Flange? _____

7. Determine distance Ⓐ . _____

8. Determine distance Ⓑ . _____

9. What is the angular dimension for the chamfer on the ⌀.39? _____

10. What type of section view is shown in the front view? _____

11. What style of thread representation is used to show the thread? _____

12. When was the last drawing change made? _____

13. Which dimension was affected by the last change? _____

14. Determine distance Ⓒ . _____

15. Determine distance Ⓓ . _____

16. What scale is used for the front view? _____

17. Determine distance Ⓔ . _____

18. What does the star symbol indicate? _____

19. What is the Flange head thickness? _____

20. What is the diameter of the hole through the center of the Flange? _____

unit 11

PIPE THREADS

Pipe threads are used for many industrial applications where a sealed joint or connection is required. The class of pipe thread is determined by its intended use. The pipe threads commonly found on industrial drawings are American National Standard Pipe Threads.

AMERICAN NATIONAL STANDARD

Two types of pipe threads are approved as American National Standards. They are tapered pipe threads and straight pipe threads. Each type is used for different applications.

Tapered Threads

Tapered pipe threads are recommended for general use. They have a taper of 0.75 inch per foot measured on the diameter along the axis. The thread angle is 60 degrees and standards for sizes and pitches have been established. The taper of the thread insures easy starting and a tight joint which will hold liquid or gases under pressure. A sealer is usually applied to pipe thread joints to prevent leakage.

Modified taper pipe threads are used in some special purpose applications:

- *Dryseal pressure-tight joints* — A metal-to-metal joint which eliminates the need for a sealer. This type joint is used in automobile, aircraft, and marine applications.

- *Rail-fitting joints* — Used where rigid rail joints are required.

The following symbols are used to designate American National Standard Taper Pipe Threads. Some of the more common uses of each are also indicated.

- *NPT* — American National Standard Taper Pipe for normal use
- *NPTR* — American National Standard Taper Pipe for rigid mechanical railing joints
- *NPTF* — Dryseal American National Standard Pipe Thread
- *PTS-SAE SHORT* — Dryseal SAE Short Taper Pipe Thread. This thread series is the same as the NPTF except it is shortened by one thread. This series is used when extra clearance is required.

Straight Threads

Straight pipe threads are parallel to the axis of the thread. The form of the thread is the same as the American National Standard Taper Pipe Thread. The number of threads per inch, thread angle, and depth of thread are all the same as the tapered version.

Straight pipe threads are used for pressure-tight joints, loose-fitting mechanical joints, and free-fitting mechanical joints. This series of thread form are used in low pressure situations only.

The following symbols are used to designate American National Standard Straight Pipe Threads.

- *NPSC* — American National Standard Straight Pipe for Pipe Couplings
- *NPSM* — American National Standard Straight Pipe for Free-fitting Mechanical Joints. Used where no internal pressures exist.
- *NPSL* — American National Standard Straight Pipe for Loose Fits with Lock Nuts. Used where the largest diameter pipe thread is required.
- *NPSH* — American National Standard Straight Pipe for Hose Couplings
- *NPSF* — Dryseal American National Standard Fuel Internal Straight Pipe Threads
- *NPSI* — Dryseal American National Standard Intermediate Internal Straight Pipe Threads. Generally used on hard or brittle materials.

REPRESENTATION OF PIPE THREADS

Pipe threads are usually represented on drawings in schematic or simplified form. Pictorial representation may be used on large thread diameters where detail is needed.

The taper of the pipe thread is not usually shown unless needed. A leader and note indicate the type of thread form required.

American National Standard Pipe Threads are specified as follows:

1. Nominal size
2. Number of threads per inch
3. Symbols for thread series and form

Figure 11-1 shows typical thread specification callouts. Table 11-1 lists pipe sizes for American National Standard Pipe Threads.

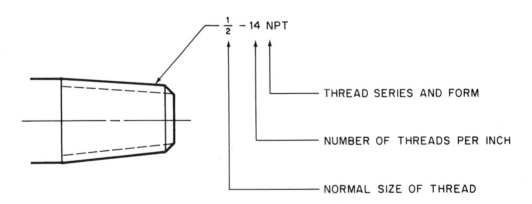

Fig. 11-1 Thread specification

Table 11-1 American Standard Pipe Threads

TAPER $\frac{3}{4}$ IN PER FT OR $\frac{1}{16}$ IN PER IN

Nominal Pipe Size	Threads per Inch	A	B	C	E	F	G	L-1	L-2	L-3	H Reamer Dia	Tap Drill Size
1/16	27	.181	.3125	.271	.301	.242	.0296	.160	.261	.390	.252	1/4
1/8	27	.269	.405	.364	.393	.334	.0296	.180	.264	.392	.345	11/32
1/4	18	.364	.540	.477	.522	.433	.0444	.200	.402	.595	.445	7/16
3/8	18	.493	.675	.612	.656	.568	.0444	.240	.408	.601	.583	19/32
1/2	14	.622	.840	.758	.816	.701	.0571	.320	.534	.782	.721	23/32
3/4	14	.824	1.050	.968	1.025	.911	.0571	.339	.546	.794	.932	15/16
1	11 1/2	1.049	1.315	1.214	1.283	1.144	.0696	.400	.683	.985	1.169	1 5/32
1 1/4	11 1/2	1.380	1.660	1.557	1.627	1.488	.0696	.420	.707	1.009	1.514	1 1/2
1 1/2	11 1/2	1.610	1.900	1.796	1.866	1.727	.0696	.420	.724	1.025	1.753	1 23/32
2	11 1/2	2.067	2.375	2.269	2.339	2.200	.0696	.436	.757	1.058	2.227	2 3/16
2 1/2	8	2.469	2.875	2.720	2.820	2.620	.100	.682	1.138	1.571	2.662	2 5/8
3	8	3.068	3.500	3.341	3.441	3.241	.100	.766	1.200	1.634	3.289	3 1/4
3 1/2	8	3.548	4.000	3.838	3.938	3.738	.100	.821	1.250	1.684	3.789	3 3/4
4	8	4.026	4.500	4.334	4.434	4.234	.100	.844	1.300	1.734	4.287	4 1/4
5	8	5.047	5.563	5.391	5.491	5.291	.100	.937	1.406	1.841	5.349	5 5/16
6	8	6.065	6.625	6.446	6.546	6.346	.100	.958	1.513	1.947	6.406	6 3/8
8	8	7.981	8.625	8.434	8.534	8.334	.100	1.063	1.713	2.147	8.400	
10	8	10.020	10.750	10.545	10.645	10.445	.100	1.210	1.925	2.359	10.521	
12	8	12.000	12.750	12.533	12.633	12.433	.100	1.360	2.125	2.559	12.518	

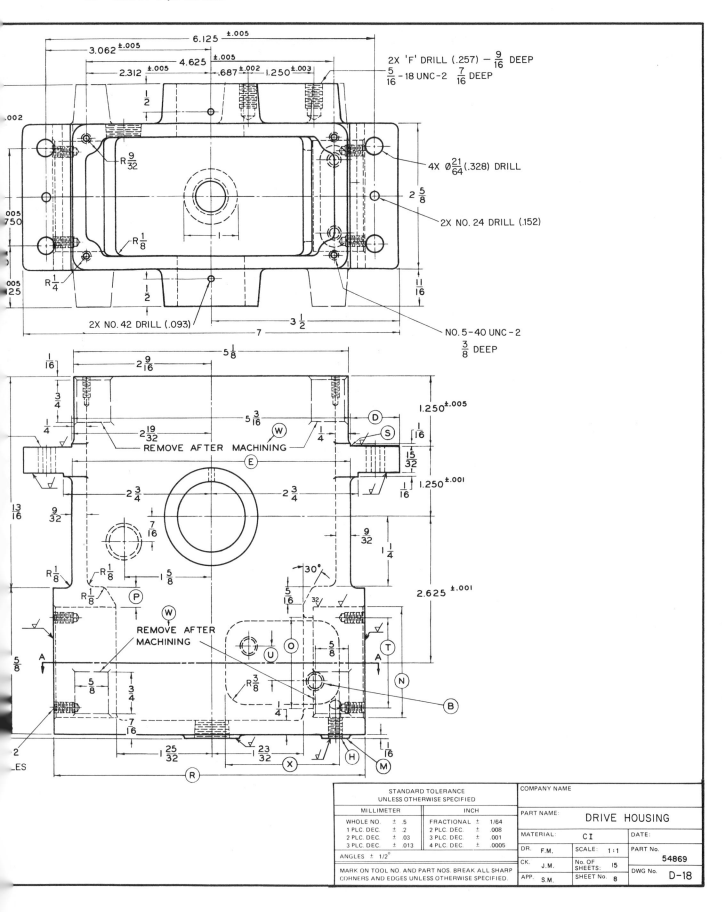

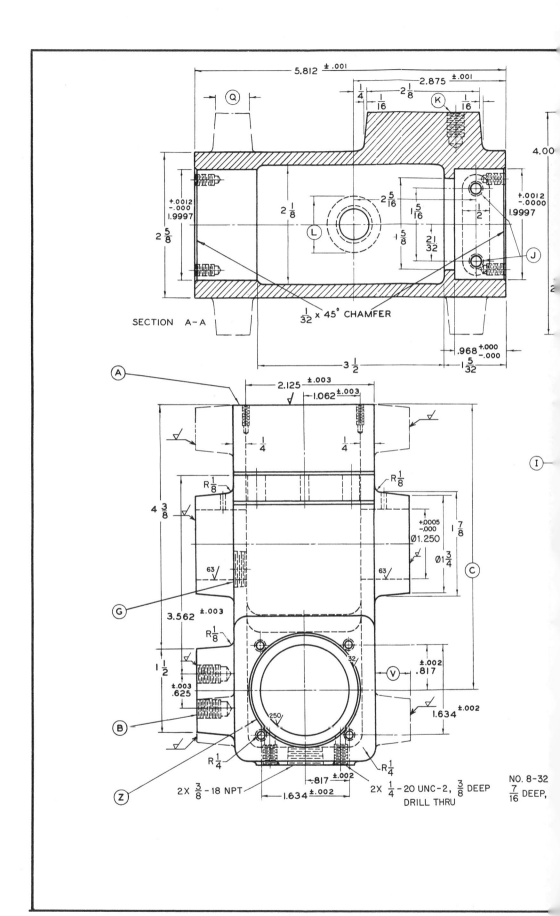

ASSIGNMENT D-18: DRIVE HOUSING

1. Determine distance Ⓣ . _____
2. Determine distance Ⓤ . _____
3. What is the finished distance on the lug at Ⓥ ? _____
4. What kind and size of hole is shown at Ⓖ ? _____
5. How many pipe-threaded holes are required? _____
6. What size pipe thread is called out? _____
7. Why are the lugs at Ⓥ shown using dash lines? _____
8. What is distance Ⓧ ? _____
9. How many temporary machining lugs are shown? _____
10. Describe the full operation for holes at Ⓑ . _____
    _____
11. What surface finish is required on the 1.9997 hole? _____
12. Do the dotted lines at Ⓘ represent a threaded hole? _____
13. Identify holes Ⓙ in the front view. _____
14. Identify hole Ⓚ in the front view. _____
15. What is the dimension at Ⓛ ? _____
16. Is section A-A taken through the top view, the front view, or the left-side view? _____
17. Determine distance Ⓒ . _____
18. What is the width of pad Ⓜ ? _____
19. What are the sizes at Ⓞ and Ⓝ ? _____
20. What are the dimensions at Ⓟ , Ⓠ , and Ⓡ ? Ⓟ = _____
    Ⓠ = _____
    Ⓡ = _____
21. What is the diameter of circle Ⓩ ? _____
22. How far from surface Ⓐ are the top edges of the upper machining lugs? _____
23. Determine distances Ⓓ and Ⓔ . Ⓓ = _____
    Ⓔ = _____
24. What is the length of the pad at Ⓜ ? _____
25. What machining operation is involved by statement Ⓦ ? _____

unit 12

IDENTIFYING STEELS

All producers and users of steel and its alloys are concerned with identifying the material according to its chemical content.

American mass production methods proceed on the basis that each manufacturer can have exactly the metal that a product requires. Metals must be of uniform quality in every shipment. For this reason, standards and numbering systems for identifying carbon steels have been established by a variety of engineering societies, trade organizations, and private industries.

The primary numbering systems most commonly accepted and used throughout the metalworking industry are those developed by the American Iron and Steel Institute (AISI) and the Society of Automotive Engineers (SAE). Both systems are very similar.

AISI AND SAE SYSTEMS

The numbering system developed by the AISI and SAE consists of four digits which are used to identify carbon and alloy steels. A fifth digit is used when identifying certain chromium steels. The first digit indicates the type of steel. The second digit indicates either the average percentage of the main alloying element (other than carbon) or the presence of a second alloying element. The last two (or three) digits indicate the average carbon content in points. A point is equal to 0.01%, Figure 12-1.

A drawing marked SAE 71360 indicates a tungsten steel of about 13 percent tungsten (12 to 15) and .60 percent carbon (.50 to .70). For a steel marked SAE 1020, the *1* indicates a carbon steel, *0* indicates that the steel is not alloyed, and the *20* indicates an average carbon content (.15 to .25) of .20 percent. The drawing of the Worm Spindle (Unit 21) indicates that the material to be used is SAE 1040.

In the list of steel types in Table 12-1, the small *x*'s at the end of the numerals indicate the places of the digits. These digits, when put in place, further subdivide each type of steel into subtypes as mentioned. Carbon steels are often referred to as low-, medium- or high-carbon steels.

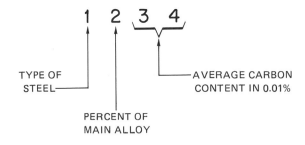

Fig. 12-1 AISI and SAE numbering systems

Table 12-1 Basic SAE Numbering System for Steels

TYPE OF STEEL	SAE NUMERALS
Carbon Steels	1xxx
Plain carbon	10xx
Free cutting (screw stock)	11xx
High manganese	13xx
Nickel Steels	2xxx
3.50% nickel	23xx
5.00% nickel	25xx
Nickel Chromium Steel	3xxx
1.25% nickel, 0.60% chromium	31xx
3.50% nickel, 1.50% chromium	33xx
Molybdenum Steels (0.25% molybdenum)	4xxx
Chromium 1.0%	41xx
Chromium 0.5%, nickel 1.8%	43xx
Nickel 2%	46xx
Nickel 3.5%	48xx
Chromium Steels	5xxx
Low chrome	51xx
Medium chrome	52xx
Chromium Vanadium Steels	6xxx
Nickel-Chromium-Molybdenum (low amounts)	8xxx
Silicon-Manganese	92xx

We must always keep in mind that the physical characteristics and chemical content of metals have a very definite relationship. Some individual subtypes of plain carbon steels (classification 10xx), with the percentage ranges of their carbon content, are given in Table 12-2. Some subtypes of free-cutting carbon steels are shown in Table 12-3.

Tables are available giving full lists of steels, their compositions, and code numbers. These tables are contained in handbooks published by the Society of Automotive Engineers, in *Machinery's Handbook*, and in trade manuals. The Tables serve as a reference when selecting material of a definite composition to meet a specific condition.

Table 12-2 SAE Specifications for Plain Carbon Steels

SAE NUMBER	CARBON RANGE PERCENT	SAE NUMBER	CARBON RANGE PERCENT	SAE NUMBER	CARBON RANGE PERCENT
1010	.08 – .13	1040	.37 – .44	1070	.65 – .75
1015	.13 – .18	1045	.43 – .50	1075	.70 – .80
1020	.18 – .23	1050	.48 – .55	1080	.75 – .88
1025	.22 – .28	1055	.50 – .60	1085	.80 – .93
1030	.28 – .34	1060	.55 – .65	1090	.85 – .98
1035	.32 – .38	1065	.60 – .70	1095	.90 – 1.03

Table 12-3 SAE Specifications for Free-cutting Carbon Steels

SAE NUMBER	CARBON RANGE PERCENT	SAE NUMBER	CARBON RANGE PERCENT	SAE NUMBER	CARBON RANGE PERCENT
1112	0.13 max.	1115	.13 – .18	1120	.18 – .23

Unit 12 Identifying Steels

Fig. 12-2 Number indicates that lead has been added to the steel

In addition to the numbers, certain letters may be inserted in the middle or at either end of the number for further description. A *B* or an *L* may be placed between the first and second pair of numbers. The *B* identifies the steel as a boron steel; the *L* indicates that lead has been added to increase machinability, Figure 12-2.

Prefix letters are used most frequently in the AISI system. The prefix letter identifies the steelmaking process used. The use of this letter is, at times, important because two steels having practically the same composition, but made by different processes, will often have slight but important differences in their properties. The letter prefixes with their meanings are shown in Figure 12-3.

Suffix letters may also be applied to add to the identification of the steel. For example, an *A, B,* or *C* may be applied to distinguish between steels which differ only in carbon content. An *F* may be used to describe the steel as free machining or the letter *H* may be applied to indicate that the steel was produced within certain hardenability limits, Figure 12-4.

Standard identification numbers are very convenient for placement on drawings and prints to specify the types of steels used. They give very specific information as to the steel requirements for a particular application.

```
A = OPEN HEARTH ALLOY STEEL
B = ACID BESSEMER CARBON STEEL
C = BASIC OPEN HEARTH CARBON STEEL
D = ACID OPEN HEARTH CARBON STEEL
E = ELECTRIC FURNACE STEEL
```

EXAMPLE:

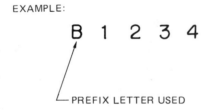

Fig. 12-3 AISI numbering system letter prefixes

Fig. 12-4 Use of suffix letter in identifying number

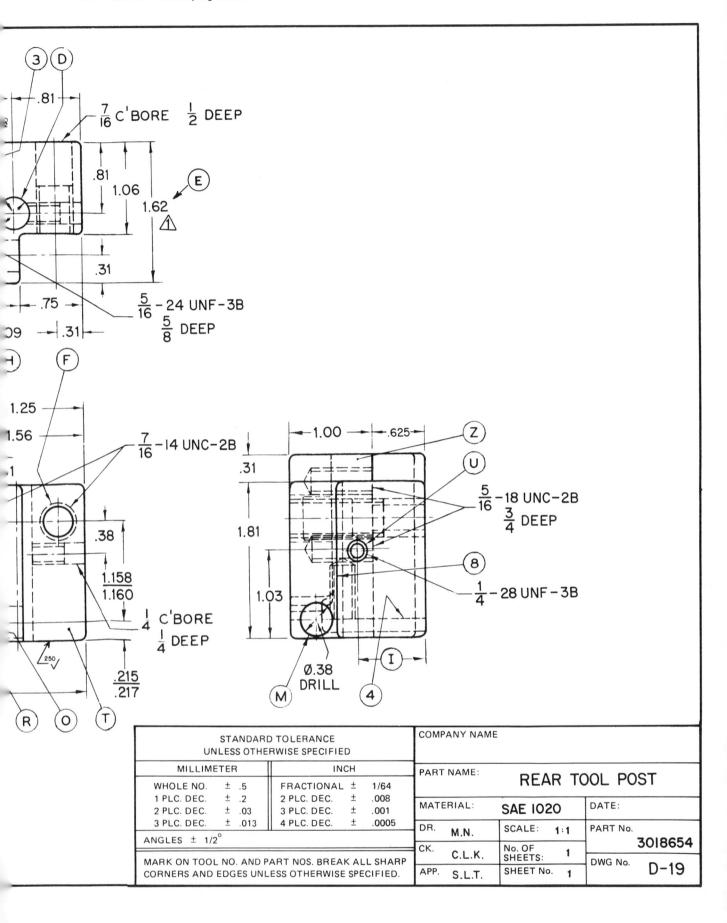

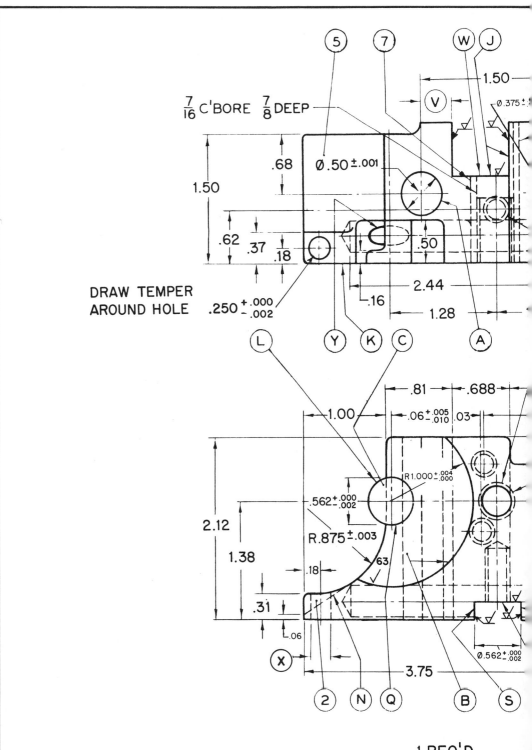

1 REQ'D
PACK HARDE

ASSIGNMENT D-19: REAR TOOL POST

1. What material is specified for the Rear Tool Post? _____
2. What is the approximate carbon content of the steel? _____
3. What does the first digit *1* indicate in the material specifications? _____
4. If the prefix *C* were added to the material specification, what would it mean? _____
5. How deep is circular slot Ⓑ ? _____
6. Which of the threaded holes are counterbored? What is the depth of each counterbore? _____
7. What is distance Ⓘ ? _____
8. What is the width and depth of slot Ⓙ ? _____
9. Indicate surface Ⓚ in the front view. _____
10. Indicate surface Ⓝ in the top view. _____
11. Locate lines Ⓢ and Ⓞ in the top view. _____
12. Locate the projection of line ④ in the front view. _____
13. Give distance between lines ⑦ and ③ . _____
14. What is the maximum allowable outside diameter of the slot Ⓑ ? _____
15. What is the diameter of the boss shown at Ⓨ ? _____
16. What line in the right-side view indicates surface Ⓣ ? _____
17. What is the depth of hole Ⓜ ? _____
18. Determine distance Ⓥ . _____
19. What is the depth of the counterbore for hole Ⓗ from surface Ⓦ ? _____
20. Locate surface ② in the top view. _____
21. What is the shape of the intersection between circular slot Ⓑ and hole Ⓜ ? In what view does the shape of this intersection appear? _____
22. How many threaded holes are indicated? _____
23. Determine the overall height of the Tool Post. _____
24. What is the depth of the threaded hole shown at Ⓡ ? _____
25. What diameter counterbore is required for the hole at Ⓤ ? _____

unit 13

DOVETAILS

DESCRIPTION OF DOVETAILS

A dovetail refers to a groove or slide whose sides are cut at an angle, making an interlocking joint between two pieces so as to resist pulling apart in all directions except along the ways of the dovetail slide itself.

The dovetail is commonly used in the design of slides on such machine parts as the cross slide of a lathe, the slide on the underside of a milling machine table, and for other sliding parts.

The two parts of a dovetail slide are shown in cross section in Figure 13-1.

When the dovetail parts are to be machined to a given width, they may be gauged by using accurately sized cylindrical rods or wires.

Dovetails are usually dimensioned as shown in Figure 13-2. The dimensions limit the boundaries to which the machinist works.

The edges of a dovetail are usually broken to remove the sharp corners. On large dovetails the external and internal corners are often machined as shown in Figure 13-3 at *A* or *B*.

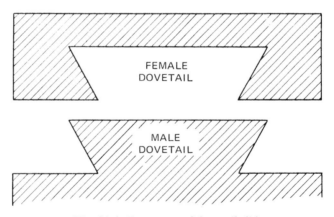

Fig. 13-1 Two parts of dovetail slide

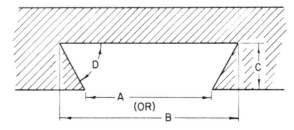

Fig. 13-2 Dimensions of a dovetail

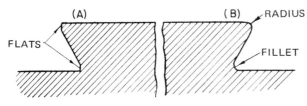

Fig. 13-3 Types of corners used on large dovetails

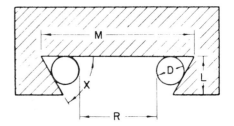

Fig. 13-4 Measuring female dovetail

MEASURING DOVETAILS

Female dovetails may be measured by placing two accurate rods of known diameter against the sides and bottom of the dovetail as shown in Figure 13-4. The distance R is then measured and checked against the computed value of R as found by the following formula:

$$R = M - \left[D \left(1 + \cot \frac{\text{ANGLE X}}{2} \right) \right]$$

NOTE: Diameter D should be slightly less than distance L.

Example:

Given:
- Distance M = 3.000″
- Diameter of Rod D = .625″
- Degrees in Angle X = 55°

Then:

$R = 3.000 - .625 (1 + 1.921)$

Combining: $R = 3.000 - 1.8256$

$R = 1.1744″$

Male dovetails may also be measured in a similar manner by placing two accurate rods of known diameter against the sides and bottom of the dovetail as shown in Figure 13-5. The distance Q over the rods is measured and then checked against the computed value of Q as found by the following formula:

$$Q = D \left(1 + \cot \frac{\text{ANGLE X}}{2} \right) + S$$

Example:

Given:
- Distance S = 2.000″
- Diameter of Rod D = .625″
- Degrees in Angle X = 55°

Then:

$Q = .625 (1 + 1.921) + 2$

Combining: $Q = (.625 \times 2.921) + 2$

$Q = 3.8256″$

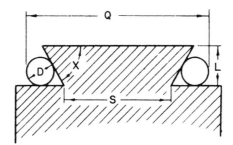

Fig. 13-5 Measuring male dovetail

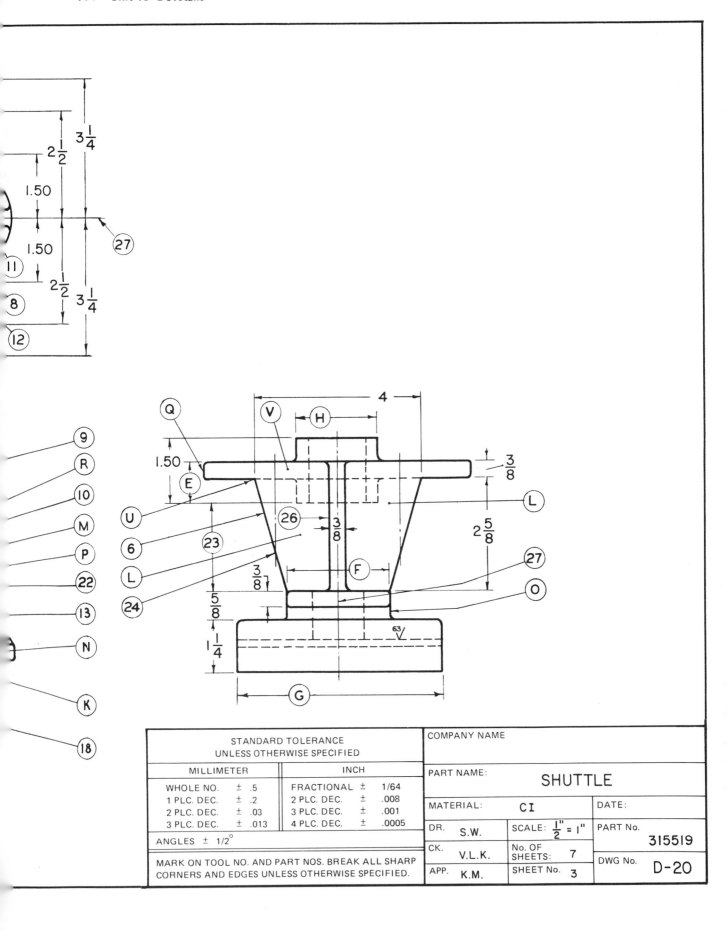

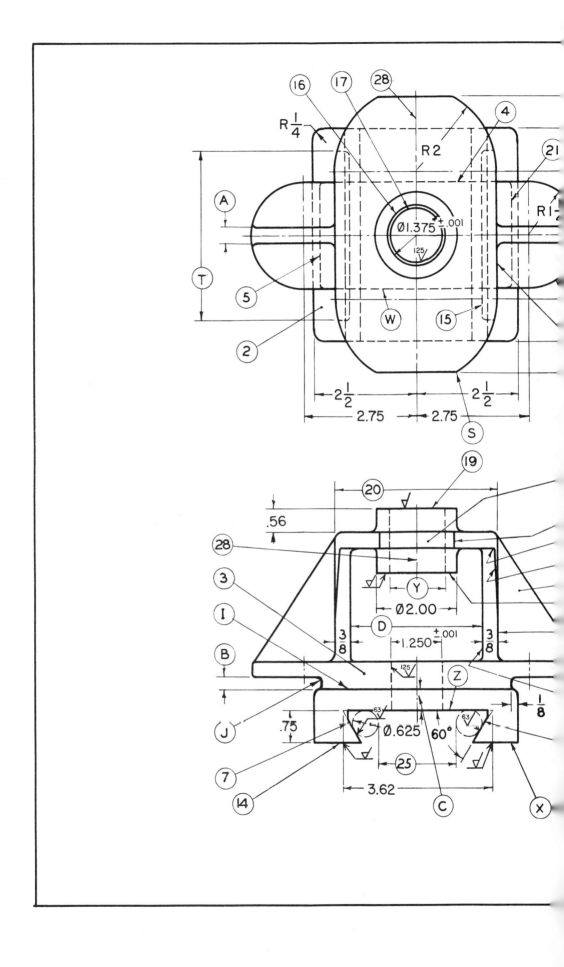

ASSIGNMENT D-20: SHUTTLE

1. Determine the following distances:

   Ⓐ _____ Ⓒ _____ Ⓔ _____ Ⓖ _____
   Ⓑ _____ Ⓓ _____ Ⓕ _____ Ⓗ _____

2. What surface in the top view does line Ⓘ represent?

3. What line or surface in the top view does line Ⓙ represent?

4. What line or surface in the top view does line Ⓚ represent?

5. What line in the right-side view represents line Ⓜ ?

6. What line or surface in the top view represents surface Ⓝ ?

7. What line or surface in the top view does line Ⓞ represent?

8. What lines in the top view and the right-side view represent surface Ⓟ ?

9. What surface in the front view does line Ⓠ represent?

10. What line in the right-side view represents line Ⓡ ?

11. What line in the right-side view represents corner Ⓢ ?

12. Determine distance Ⓣ .

13. What point in the front view shows point Ⓤ ?

14. Determine distance Ⓨ .

15. What surface represents line Ⓦ in the front view?

16. What line in the top view represents surface Ⓥ ?

17. What is the overall height of the Shuttle?

18. Determine distance ⃝20 .

19. What is the extreme length of the Shuttle?

20. What is the distance between lines ⃝5 and ⃝21 ?

21. Determine distance ⃝23 .

22. What line in the front view represents surface Ⓛ ?

23. Determine dimension ⃝25 when measuring rods are .625 diameter.

Ⓐ = _____
Ⓑ = _____
Ⓒ = _____
Ⓓ = _____
Ⓔ = _____
Ⓕ = _____
Ⓖ = _____
Ⓗ = _____

Unit 13 Dovetails 115

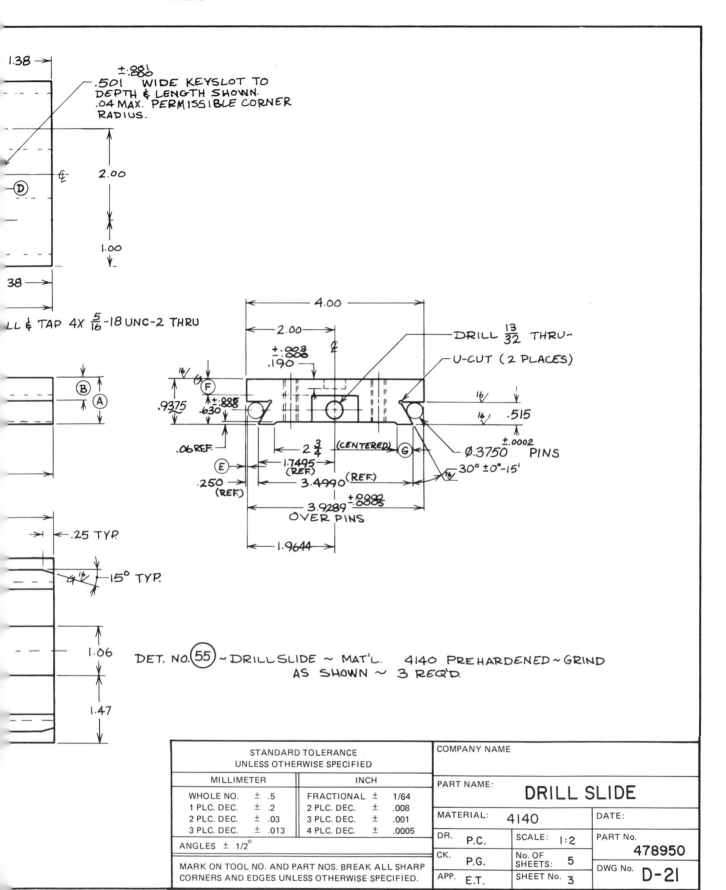

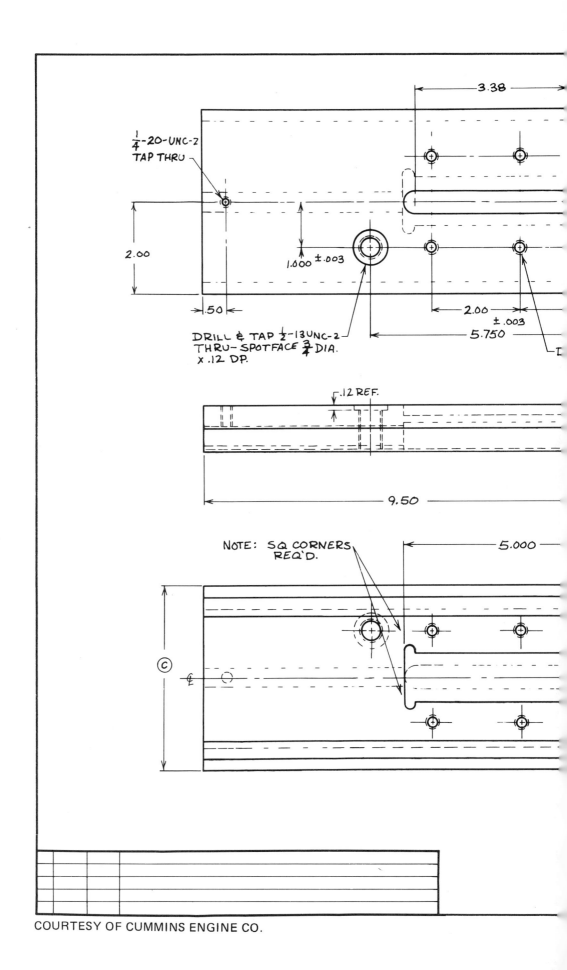

COURTESY OF CUMMINS ENGINE CO.

ASSIGNMENT D-21: DRILL SLIDE

1. What material is required for the Drill Slide?   _____
2. Determine dimension Ⓐ.   _____
3. Determine dimension Ⓑ.   _____
4. What is the angular dimension of the dovetail?   _____
5. What four views of the drill slide are shown?   _____

   _____

   _____

   _____

   _____

6. Determine dimension Ⓒ.   _____
7. What tolerance is allowed on the dovetail angle?   _____
8. What is the width of the dovetail as measured over
   the pins?   _____
9. What diameter pins are used for the dovetail mea-
   surement?   _____
10. What machining process is required to finish the top
    and bottom of the Slide?   _____
11. Determine dimension Ⓓ.   _____
12. Determine dimension Ⓔ.   _____
13. What is the overall length of the Slide?   _____
14. What is the depth of the .501-wide keyslot?   _____
15. What is the length of the slot in the bottom of the
    Slide?   _____
16. What is the depth of the slot?   _____
17. Determine dimension Ⓕ.   _____
18. Determine dimension Ⓖ.   _____
19. What is the depth of the 3/4-diameter spotface?   _____
20. How many spotfaced holes are required?   _____

unit 14

CASTING

SAND MOLDING

Irregular or odd-shaped parts which would be difficult to make from metal plate or bar stock may be cast to the desired shape in sand molds. A pattern may be a wood or metal model of the part so constructed that it may be used to form the impression, called a mold, in the sand. Molten metal, when poured into this mold and solidified, will form a casting of the desired part.

The sand for the mold is held in place in a wood or metal frame, or box, called a *flask*. The assembly of the entire flask is shown in Figure 14-1. The opened flask is illustrated in Figure 14-2. Figure 14-3 shows a simple pattern, and Figure 14-4 shows it as it is used to form the mold.

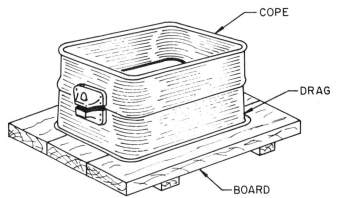

Fig. 14-1 Assembled flask

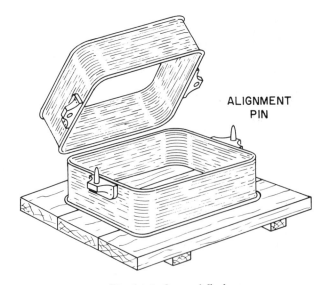

Fig. 14-2 Opened flask

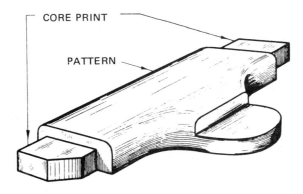

Fig. 14-3 Wood pattern

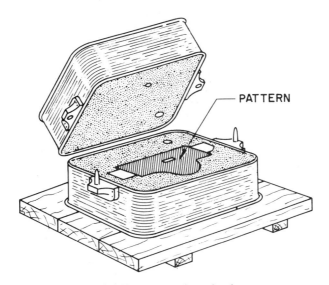

Fig. 14-4 Pattern ready to be drawn

In Figure 14-5, the pattern (A), is removed and the required core is set in place. The flask is then closed, Figure 14-6, and the mold is ready to be poured. A heavy weight is placed on top of the flask to retain the metal during the pouring process. The hot metal poured into the *sprue* at (A) flows along a channel called the *gate* (B) into the cavity or mold made by the removal of the pattern (C).

The resultant casting, except for the attached sprue, gate, and riser, as shown in Figure 14-7, is of the same shape as the original pattern and is slightly smaller in size due to the natural shrinkage of a cooling body, for which allowance has been made on the pattern.

FLAT BACK PATTERNS

Simple shapes such as the one shown in Figure 14-8A are very easy to mold. In this case the flat face of the pattern is at the parting line and lies perfectly flat on the molding board. In this position, no molding sand sifts under the flat surface to interfere with the drawing of the pattern. The simplicity with which flat back patterns of this type may be drawn from a mold is illustrated in Figure 14-8B.

120 Unit 14 Casting

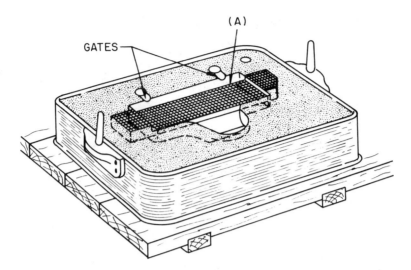

Fig. 14-5 Mold with core set in place

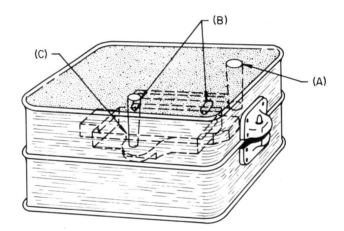

Fig. 14-6 Mold closed

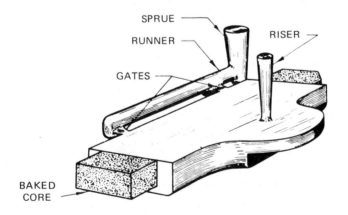

Fig. 14-7 Casting as removed from the mold

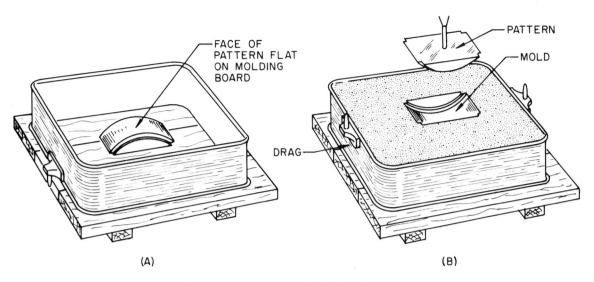

Fig. 14-8 Making a mold of a flat back pattern

CORING

When casting a recess or odd-shaped hole or cavity in a part, it is often necessary to make a separate mold of the cavity to the exact shape in a box, known as a core box, Figure 14-9. The core box is then filled with core sand mixed with a bonding agent. The core material is rammed so that, when removed from the box, it retains the shape of the cavity, Figure 14-10. It is then baked hard in a core oven.

After the pattern has been removed and the baked core has been put in place, the core is solidly supported in the mold of the core print, permitting only that part of the core that corresponds to the shape of the cavity in the casting to project into the mold, Figure 14-5 at (A).

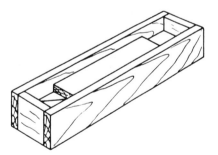

Fig. 14-9 Core box

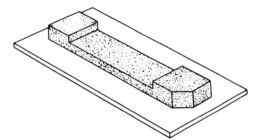

Fig. 14-10 Core ready for baking

Fig. 14-11 Completed casting with sprue, gate and riser removed and with the baked core broken out

When the molding flask is closed and the metal is poured, it flows around this core and completely fills the mold. After the metal solidifies and is removed from the mold as a casting, Figure 14-7, the baked core is broken out, leaving a cavity of the desired shape and size in the casting as shown in Figure 14-11 at (A).

Coring should be thoroughly studied by the drafter as a basis for the knowledge needed for economical construction of parts which must be cast with a hole or cavity. It is difficult to make a mold from a pattern which has been incorrectly constructed.

The drafter makes the drawing of the part to the proper shape. He also indicates the scale and gives the dimensions from which the pattern is constructed. Therefore, it is essential that he have a knowledge of the principles of molding practice and pattern making to insure a practical design.

The mechanic must also understand the processes which are involved in pattern making and molding practice that are given or implied on a drawing.

CORED CASTINGS

Cored castings have certain advantages over solid castings. Where practical, castings are designed with cored holes or openings for purposes of economy, appearance, and accessibility to interior surfaces.

The appearance of a casting is often improved by cored openings. In most instances, cored castings are more economical than solid castings due to the metal saved. While cored castings are lighter, they are so designed that strength is not sacrificed. Unnecessary machining is also eliminated because of the openings that are cast in the part.

Hand holes may also be formed by coring in order to provide an opening through which the interior of the casting may be reached. These openings also permit machining an otherwise inaccessible surface of the part, as shown in Figure 14-12.

In the case of the Auxiliary Pump Base, Assignment D-22, the body of the casting is cored and provided with openings. Some of the reasons for coring this particular part are

1. Economy: the casting is lightened to save metal and weight.
2. Accessibility to interior
    a. The coring provides openings so that the four legs on the bottom of the base may be drilled and spotfaced on the inside.
    b. The hand holes also provide accessibility for reaching into the casting.
3. Appearance: the object is designed to look more attractive.

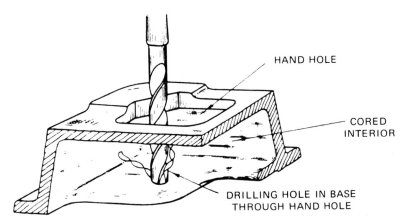

Fig. 14-12 Section of cored casting

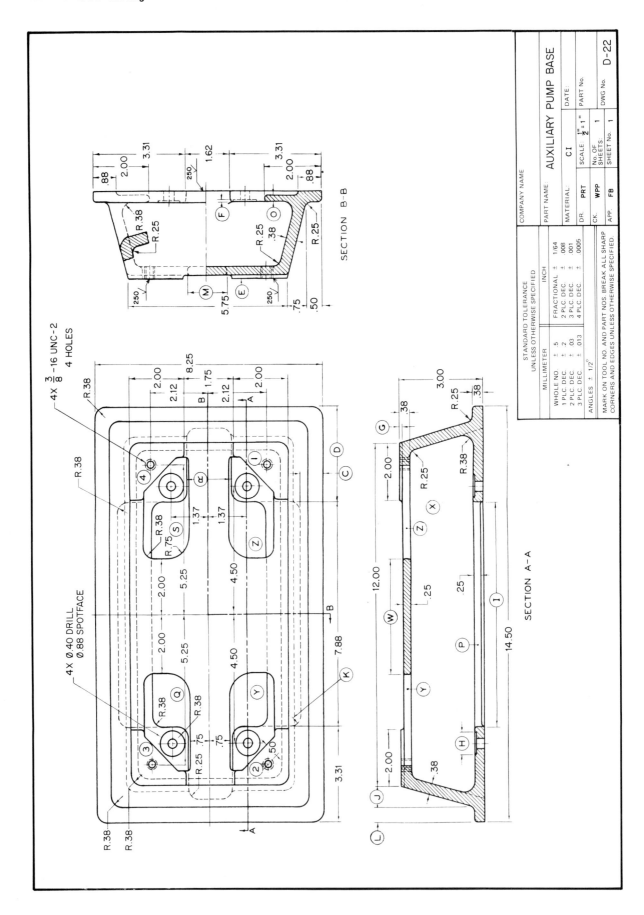

ASSIGNMENT D-22: AUXILIARY PUMP BASE

1. What surface finish is required on the pads at ①, ②, ③, ④ ?

2. What type of section view is shown at B-B?

3. What circled letters on the drawing indicate the spaces that were cored when the casting was made?

4. Locate the surface in the top view that is represented by line Ⓔ.

5. Give distance Ⓒ.

6. Give the approximate distance for Ⓕ.

7. Give the thickness of the pads at Ⓖ.

8. What is the minimum diameter for spotface Ⓗ ?

9. What is distance Ⓙ ?

10. What radius fillet would be used at Ⓚ ?

11. What is distance Ⓛ ?

12. What is distance Ⓜ ?

13. What is distance Ⓡ ?

14. What is the height of the cored area Ⓧ ?

15. What is the approximate length of the large cored area Ⓧ ?

16. What is the horizontal length of pads ①, ②, ③, and ④ ?

17. What is distance Ⓦ ?

18. What is distance Ⓘ ?

19. What is distance Ⓞ ?

20. What reason might be given for providing the openings Ⓠ, Ⓢ, Ⓨ, and Ⓩ ?

unit 15

FINISHES AND PROTECTIVE COATINGS

Machined parts are frequently given special surface treatments to protect them or increase the efficiency of their operation. The type of finish or protective coating applied depends upon how the part is to be used or the elements to which it may be exposed. Surface finishes provide one or more characteristics:

- corrosion resistance
- lubrication quality
- wear resistance
- prepaint surface finish
- improve overall appearance

The most common types of protective finishes are chemically, electrically, or physically applied. There are a wide variety of surface treatments for different types of metals. The finish may be applied before any machining is done, between the various stages of machining, or after the part has been completed. The information concerning the type of finish is usually specified on the drawing in a special notation.

Typical surface treatments for metal parts include conversion coating, electroplating, flame spray coating, and organic coating.

CONVERSION COATING

Conversion coating involves a chemical reaction between the part and protective finish material. The result of the controlled chemical attack is a bonding of materials on the surface of the part. The bonded elements improve or enhance the surface quality of the metal. Steel or iron parts are commonly given phosphate treatments as a base for paint, or to improve lubrication, wear resistance, or corrosion resistance. The phosphate treatment may be iron, zinc, or manganese compositions, Figure 15-1.

Iron phosphate is used as a base material to which a painted surface will be applied. Iron phosphate treatment is used on parts with indoor applications.

Zinc phosphates are used when finish painted parts are to be used outdoors, or when lubrication quality must be increased. Zinc coatings may also be used where limited wear resistance is required.

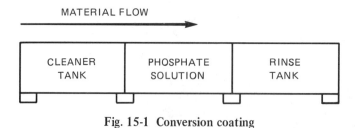

Fig. 15-1 Conversion coating

Manganese phosphate treatment etches the base metal, providing lubrication and wear resistance qualities. The black finish which results from the process also tends to improve the appearance and corrosion resistance.

Treatments for other types of metals are also commonly applied. Aluminum is usually anodized to form a protective coating while copper alloys may be bright dipped or buffed.

ELECTROPLATING

Electroplating, as the name implies, is a process which involves applying protective coatings electrically. The plating is accomplished by submerging the part in a solution containing particles of the plating material. A direct current is then passed from a power source to the part, causing the plating material to be deposited on the part.

Electroplating improves the appearance and protects parts against corrosion. Plating may also be desired where worn surfaces must be built up. Automobile trim is an example of a chrome plating process.

FLAME SPRAY COATING

Flame spray coating applies a metal or ceramic finish to machined parts. The process usually involves the salvage of worn surfaces or is applied to areas subject to excessive wear.

In flame spraying, the coating material is fed into an extremely hot flame where it is melted at approximately 5000°F. The material is then atomized with compressed air and blown against the surface to be plated. As the material cools and hardens, it forms the protective finish.

ORGANIC COATINGS

A variety of organic coatings may be applied to surfaces to protect against corrosion or to improve appearance. Examples of organic coatings include vinyl, acrylic latex, and epoxys. Most applications are made with a brush or by spraying the material onto the part.

Unit 15 Finishes and Protective Coatings

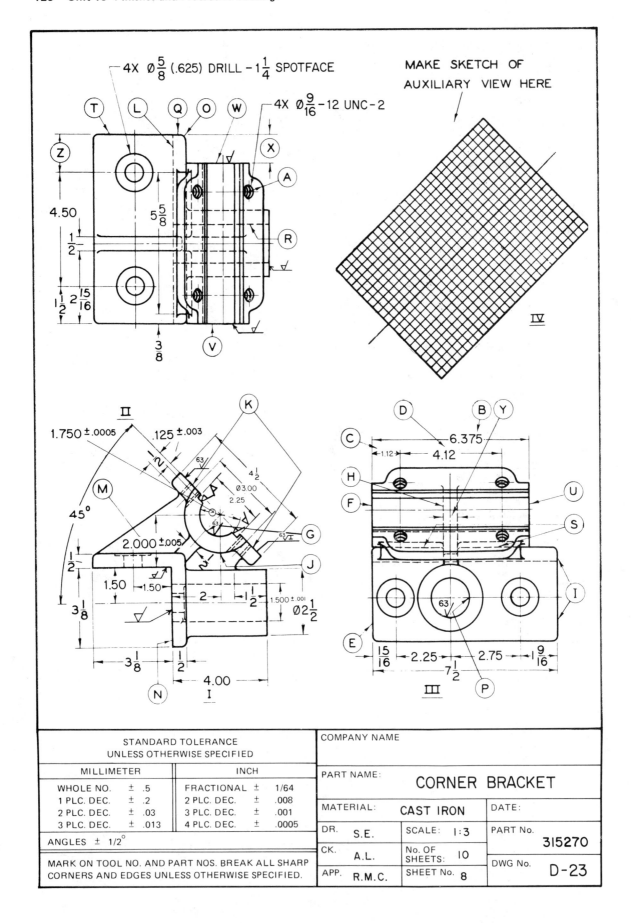

ASSIGNMENT D-23: CORNER BRACKET

1. Make a freehand working sketch in the space provided at IV on the drawing of the Corner Bracket, showing an auxiliary view of the split bearing surfaces Ⓚ and Ⓖ.
2. What surface finish is required on the 1.500 ± .001 hole?
3. Indicate the number and size of the threaded holes.
4. Indicate the size and number of holes bored.
5. Indicate the size and number of holes drilled.
6. Determine dimension Ⓐ and place it correctly on the sketch of the auxiliary view.
7. Place dimensions Ⓑ, Ⓒ, and Ⓓ correctly on the sketch of the auxiliary view.
8. Note that surface Ⓔ is not finished, but surface Ⓕ is to be finished. Allowing 1/8" for finishing, how long will the rough casting be in view III?
9. What line in view III shows surface Ⓜ?
10. In which view, and by what line, is the surface represented by line Ⓝ shown?
11. What line or surface in view III shows the projection of point Ⓙ?
12. What point or surface in view III does line Ⓡ represent?
13. Locate surface Ⓘ in view II.
14. Locate surface Ⓤ in view II.
15. Locate Ⓥ in view III.
16. What is dimension Ⓨ?
17. What is dimension Ⓩ?
18. Determine dimension Ⓧ.
19. What protective finish requirement is called out?
20. What does the lay sumbol indicate on the inclined surface?

unit 16

STRUCTURAL STEEL SHAPES

COMMON SHAPES

A variety of standard steel shapes is commercially available for use in fabricating machine parts. The structural shapes are formed by hot-rolling steel billets at extremely high temperatures. The use of these preformed shapes reduces material and machining costs involved in design work. Some of the common shapes to which structural steel is rolled include squares, flats, rounds, plates, angles, beams, and channels. Also available is a variety of steel piping and tubing. Figure 16-1 shows the cross sections of various steel shapes.

SHAPE DESIGNATIONS

The American Iron and Steel Institute (AISI) and the American Institute of Steel Construction (AISC) have jointly developed standard designations for hot-rolled structural steel

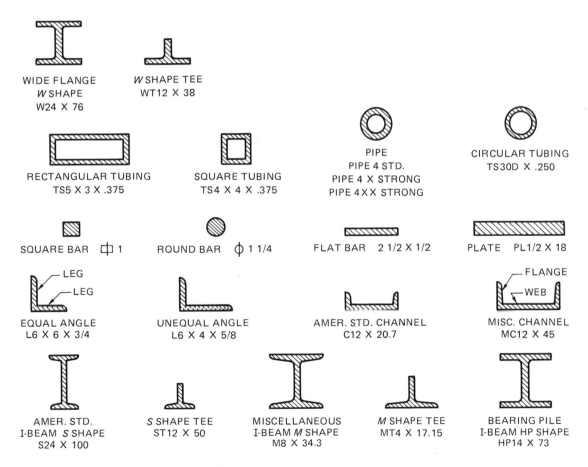

Fig. 16-1 Structural steel shapes

131

Table 16-1 AISI and AISC Hot-rolled Structural Steel Shape Designations

Designation	Type of Shape
W 24 × 76 W 14 × 26	W shape
S 24 × 100	S shape
M 8 × 18.5 M 10 × 9 M 8 × 34.3	M shape
C 12 × 20.7	American Standard Channel
MC 12 × 45 MC 12 × 10.6	Miscellaneous Channel
HP 14 × 73	HP shape
L 6 × 6 × $\frac{3}{4}$	Equal Leg Angle
L 6 × 4 × $\frac{5}{8}$	Unequal Leg Angle
WT 12 × 38 WT 7 × 13	Structural Tee cut from W shape
ST 12 × 50	Structural Tee cut from S shape
MT 4 × 9.25 MT 5 × 4.5 MT 4 × 17.15	Structural Tee cut from M shape
PL $\frac{1}{2}$ × 18	Plate
Bar 1	Square Bar
Bar $1\frac{1}{4}$	Round Bar
Bar $2\frac{1}{2}$ × $\frac{1}{2}$	Flat Bar
Pipe 4 Std. Pipe 4 X - Strong Pipe 4 XX - Strong	Pipe
TS 4 × 4 × .375 TS 5 × 3 × .375 TS 3 OD × .250	Structural Tubing: Square Structural Tubing: Rectangular Structural Tubing: Circular

shapes, Table 16-1. The designations are used by the steel producing and fabricating industries when designing or ordering materials. The identification system provides information as to the cross-sectional shape, size, and weight of the member. For example, the section symbol for an I-beam is the letter *S*, which denotes shape, followed by the actual depth in inches and weight in pounds per foot. Some examples are provided in Figure 16-2. Angle structural shapes may have equal or unequal legs. The section symbol for an angle is *L*, followed by the length of each leg and the thickness of the angle.

All structural shapes come in a variety of sizes and have different thicknesses for the same leg lengths. Handbooks on steel, such as those issued by the American Institute of Steel Construction, are a good reference for data on shapes, weights, and sizes.

A sample identification for a plate reads

PL 5/8 x 18

meaning the plate is 18" wide and 5/8" thick.

A sample identification for an I-beam reads

S15 x 75.0

meaning that the I-beam is 15" deep and weighs 75 pounds per foot.

A sample identification for a channel reads

C9 x 25.0

meaning that the standard channel is 9" high and weighs 25 pounds per foot.

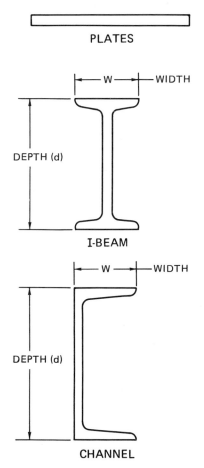

Fig. 16-2 Identifying structural steel shapes

BUILT-UP SECTIONS

Various combinations of structural steel shapes may be used to form built-up sections. The units may be riveted or welded together to form an assembly. The print reader must be familiar with the conventional symbols used to represent the shapes for built-up sections. Figure 16-3 shows some typical examples of built-up sections.

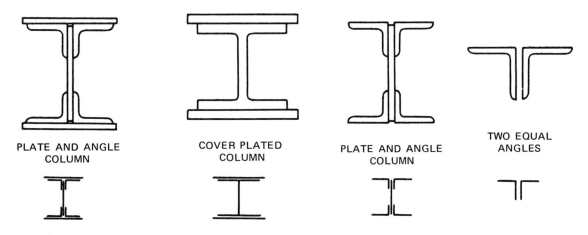

Fig. 16-3 Built-up sections

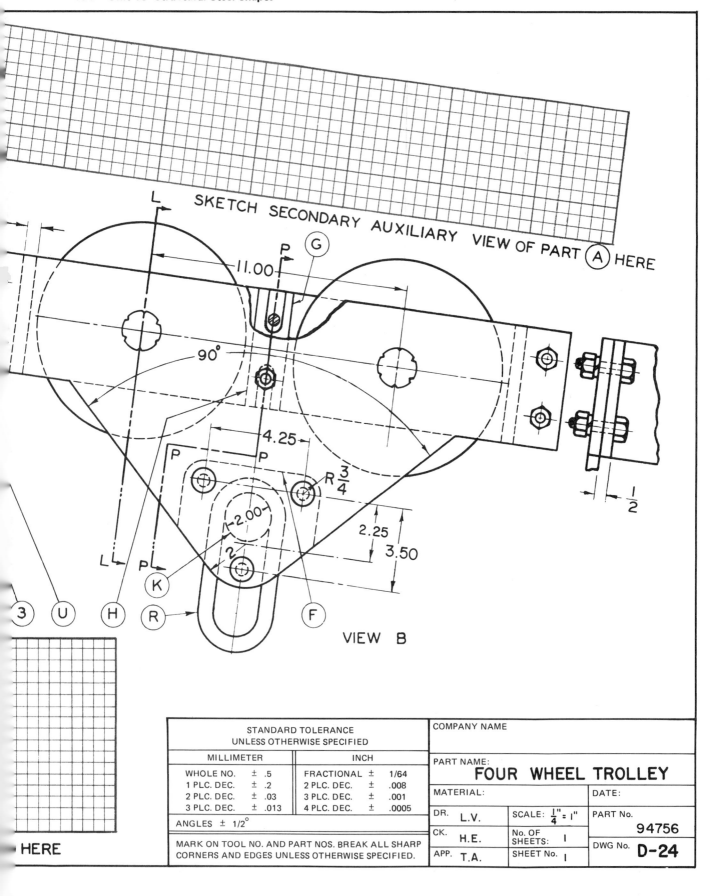

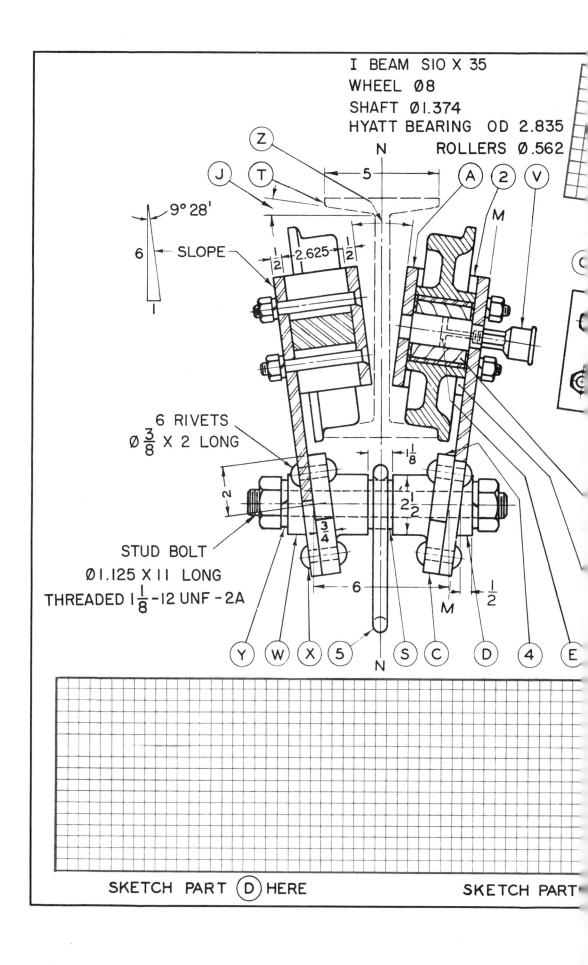

ASSIGNMENT D-24: FOUR-WHEEL TROLLEY

NOTE: Standard mechanical accessories are sometimes incorporated in the design of machine parts for economical production. These accessories are either specified on the print according to manufacturer's specifications and descriptions or to established handbook standards. The use of a handbook and manufacturers' catalogs is essential for determining detailing standards, characteristics of a special part, methods of representing, and so on.

The drawing of the Four-wheel Trolley includes a few standard parts in the completed assembly. For example, note that grease cups, lock washers, Hyatt roller bearings, rivets, and nuts, all of which are accessories, are available.

On the other hand, the special countersunk head bolts and the taper washers (commonly called a *dutchman*) are not standard accessories and must, therefore, be made especially for this particular assembly.

1. Sketch a secondary auxiliary view of part Ⓐ in the space provided above the primary auxiliary view Ⓑ on the drawing of the Four-wheel Trolley.
2. Make a dimensioned working sketch of part Ⓒ in the space provided on the drawing of the Four-wheel Trolley.
3. Make a dimensioned working sketch of part Ⓓ in the space provided on the drawing of the Four-wheel Trolley.
4. What do the dotted lines at Ⓔ indicate?
5. What cutting plane line in view Ⓑ indicates where the section to the left of line N-N is taken, or could be taken?
6. What cutting plane line in view Ⓑ indicates where the section to the right of line N-N is taken, or could be taken?
7. Locate part ② in the front view.
8. What line or surface represents surface ③ in the front view?
9. What line or surface represents line ④ in the front view?
10. What is the name of part Ⓣ ?
11. What is the name of part Ⓤ ?
12. What is the name of part Ⓥ ?
13. What is the name of part Ⓦ ?
14. What is the name of part Ⓧ ?
15. What is the name of part Ⓨ ?
16. Determine angle Ⓙ .
17. Determine angle Ⓩ .
18. What is dimension Ⓞ ?
19. What is the outside diameter of part Ⓨ ? (Refer to the proper table in this text or obtain the data from *Machinery's Handbook*.)
20. What is the depth of the beam on which the Trolley rides?

unit 17

WELDING

Welding is a process used for joining parts permanently together. It often takes the place of common fastening devices such as nuts, bolts, screws, and rivets. Welding is used extensively in fabrication work. *Fabrication* is the construction of an assembly by fastening separate units together. This is often done to produce a structure which normally would need to be cast. Fabricating is a less expensive means of construction.

WELDING JOINTS

The position of the parts being welded determines the type of welding joint formed. There are five basic types of welded joints, Figure 17-1.

- Butt joint
- Corner joint
- Tee joint
- Lap joint
- Edge joint

TYPES OF WELDS

There are a variety of types of welds which may appear on a print. The selection of a particular weld depends on the joint, material thickness, strength desired, or required penetration. The physical shape of a weld is used to give each weld its name. Figure 17-2 shows some of the basic welds used to join metals.

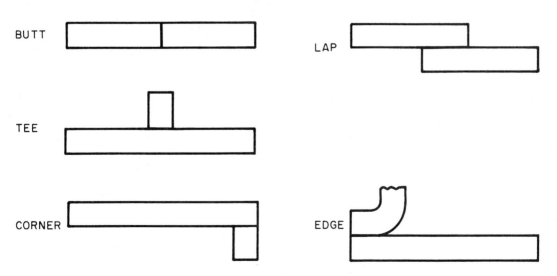

Fig. 17-1 Types of joints

137

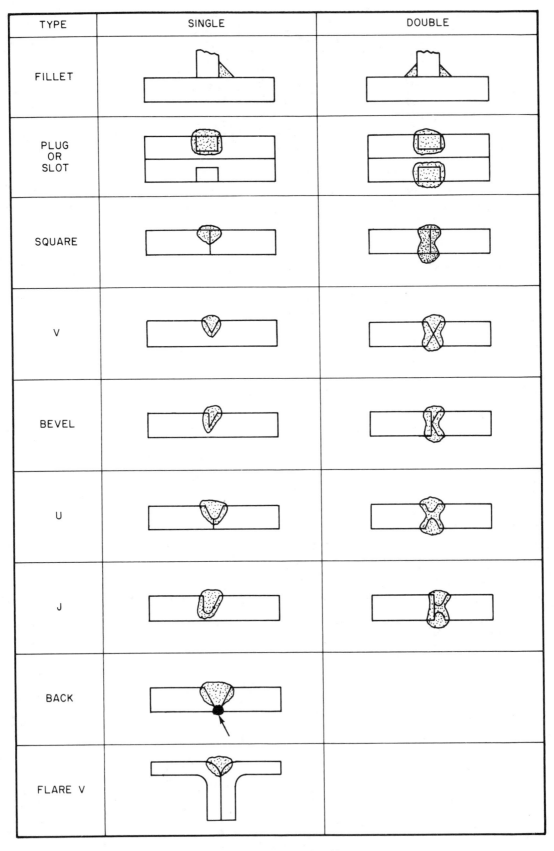

Fig. 17-2 Types of welds

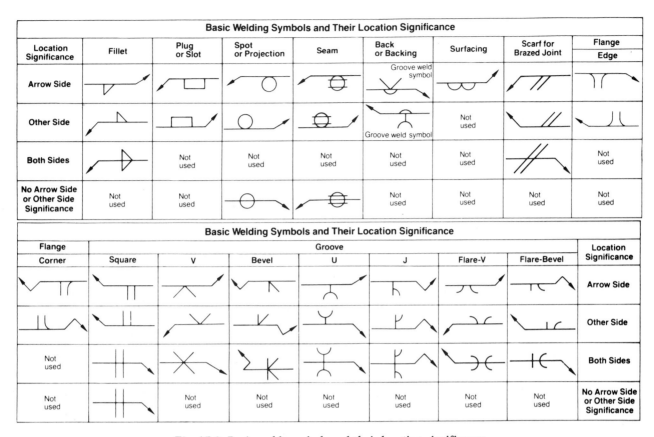

Fig. 17-3 Basic weld symbols and their location significance

WELD AND WELDING SYMBOLS

The standard graphical symbols used to convey welding information were developed by the American National Standards Institute (ANSI Y32.3-1969) and the American Welding Society (AWS A2.4-1979). The symbols are a shorthand method of transmitting information from the drafter to the welder.

The ANSI standard makes a distinction between weld symbols and welding symbols. A *weld symbol* is used to identify the type of weld required. Figure 17-3 shows the basic weld symbols used in industry.

Welding symbols may be made up of several elements of information. The information provides the specific instructions about the type, size, and location of the weld. The elements which may appear on a welding symbol are shown in Figure 17-4.

TERMINOLOGY

Reference Line — A heavy solid line which forms the body of the welding symbol. All other information is placed in positions around the reference line.

Arrow — The arrow is attached to the end of the reference line and contacts the weld joint. Welded joints are thus referred to as *arrow side* welds or *other side* welds.

Basic Weld Symbols — As previously indicated, these specify the type of weld, Figure 17-3.

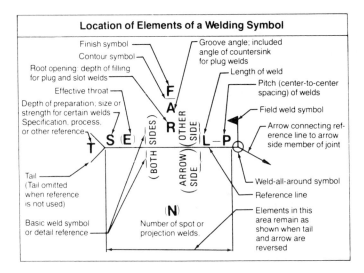

Fig. 17-4 Location of elements of a welding symbol

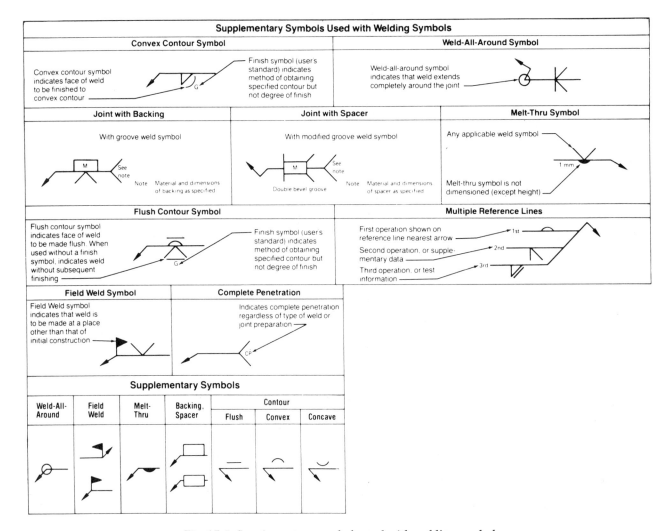

Fig. 17-5 Supplementary symbols used with welding symbols

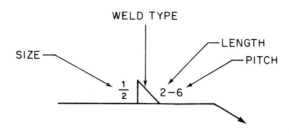

Fig. 17-6 Dimensioning a fillet weld

Supplementary Symbols — Used to provide additional information as to the extent of welding, place of welding, and bead contour. Figure 17-5 shows the supplementary symbols of the American Welding Society.

Tail — The tail appears on the end of the reference line opposite the arrow. It is used only when a specific welding process is to be specified.

Dimensions — Dimensions of a weld may specify size, length, or spacing of welds. These dimensions appear on the same side of the reference line as the weld symbol. Common practice is to call out the weld size, type, length, and center-to-center spacing (pitch), Figure 17-6.

Finish — Finish requirements may be specified below the arrow side contour symbol or above the other side contour symbol.

Process Specifications — Provided within the tail opening. This information is only specified when necessary. If the welding process is indicated elsewhere on the drawing or the specifications are known, the tail and reference are omitted from the welding symbol.

LOCATION OF WELDING SYMBOLS

The welding symbol may be placed on any of the orthographic views. It will generally be shown on the view which best shows the joint. The location of the welding symbol on each view is illustrated in Figure 17-7. However, when it is shown on one view, it is not necessary to include it on any of the other views. In this case, note that the front view is the best view for adding the symbol.

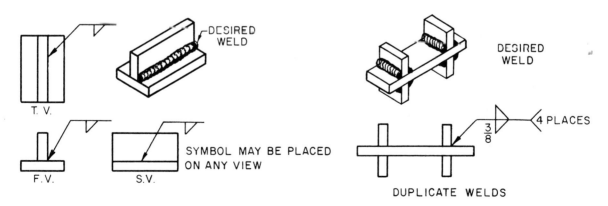

Fig. 17-7 Location of the weld symbol on orthographic views

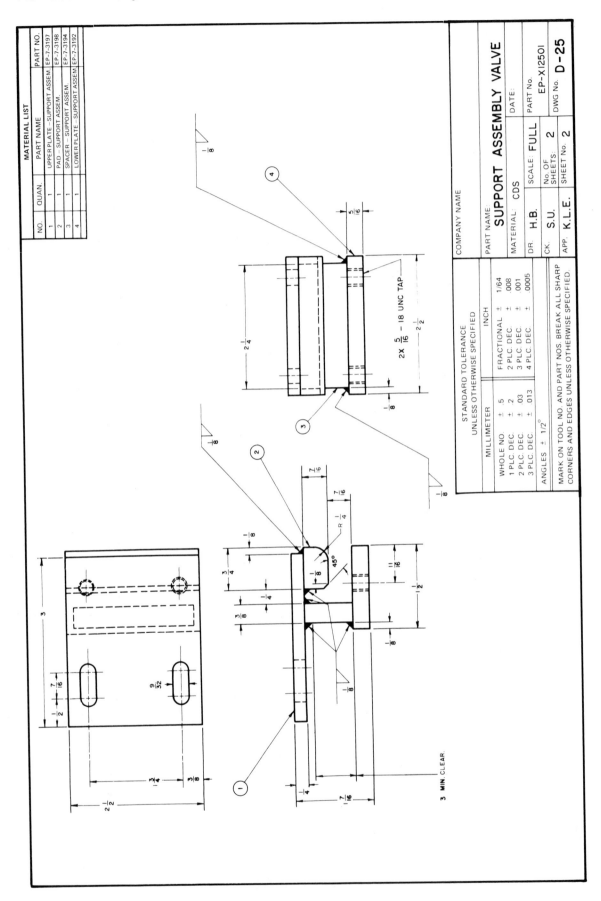

ASSIGNMENT D-25: SUPPORT ASSEMBLY VALVE

1. This small subassembly has its own name. What is it?
2. How many different pieces make up this assembly?
3. How wide is detail ①?
4. How far is it from the front edge of detail ① to the centerline of the elongated slot nearest to the front edge?
5. What is the center-to-center distance between these elongated slots?
6. How wide are these slots?
7. What is the center-to-center distance of the curved ends of these slots?
8. How far is it from the left end of detail ① to the left-hand centerline of the curved end of the slot?
9. How far is it from the right-hand edge of detail ④ to the centerlines of the tapped holes?

You may have to refer to the American Welding Society welding symbols, Figures 17-3, 17-4, and 17-5, to answer the following questions.

10. What type of welds are called for by the symbols used on the drawing?
11. How big are the welds to be that are called for on this assembly?
12. What type of information may be found in the tail of the weld symbol?
13. The line between the tail of the arrow and the arrowhead has a name. What is it?
14. What does a clear circle shown at the bend point of the reference line indicate?
15. What does a filled-in circle at this point indicate?
16. What type of weld is indicated by the little triangle attached to the reference line?
17. What size radius is used on the lower right-hand edge of detail ②?
18. How large a bevel is cut on the lower left-hand corner of detail ② and at what degree is it cut?
19. How much space is required between details ② and ③?
20. How far does detail ② extend beyond the right-hand edge of detail ①?
21. How far is it from the left-hand edge, in the front view of detail ④, to the left-hand side of detail ③?
22. The right-hand side view shows that detail ③ is not the same width as details ① and ④. How far is it from the right-hand edge of details ① and ④ to the right-hand edge of detail ③?
23. What is the minimum clearance allowed between the top surface of detail ④ and the bottom of the weld joining details ① and ③?
24. What is the total height of the assembled unit?
25. What height must be maintained between the top surface of detail ④ and the bottom of detail ②?

Unit 17 Welding

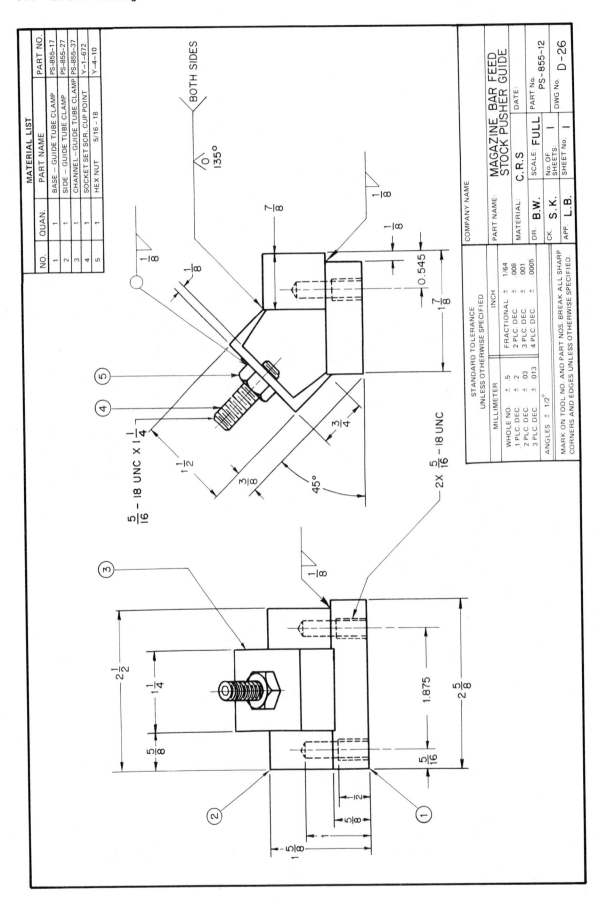

ASSIGNMENT D-26: STOCK PUSHER GUIDE

1. What is the name of the object of which this unit is a part?
2. What is the name of this specific assembly?
3. What is the name of detail ③ ?
4. How long is detail ④ ?
5. What is the diameter of detail ④ ?
6. How many threads per inch does detail ④ have?

Several welding symbols are shown on the print. The figure beside the symbol tells the size of the weld. The letters in the tail of the weld indicate a specific process or other reference. In this particular case, MA means manual arc and the figure 1 indicates that the material being welded is cold-rolled steel. For answers to questions on all other parts of the welding symbols, refer to the charts from the American Welding Society, Figures 17-3, 17-4 and 17-5.

7. How is the nut, detail ⑤ , secured to detail ③ ?
8. Explain this symbol: . What does the small circle at the joint of the symbol mean?
9. How big is this weld to be?
10. What kind of weld is called for by the small triangle on the underside of the welding symbol?
11. With what type of weld is detail ③ fastened to details ① and ② ?
12. Is any size given for this weld?
13. Is this weld used on one or both sides?
14. How are details ① , ② , and ③ fastened together?
15. What kind of nut is called for by detail ⑤ ?
16. What is the width, length, and thickness of detail ① ?
17. What is the width, length, and thickness of detail ② ?
18. What is the width, length, and depth of detail ③ ?
19. How many tapped holes are called for?
20. How deep is the threaded portion of the hole?
21. What is the total depth that the tap drill enters the work?
22. What is the center-to-center distance of the two tapped holes?
23. The back of detail ② is not even, or flush, with detail ① . How much is the offset?
24. Detail ③ is mounted at an angle to details ① and ② . At what degree is it mounted?
25. How far is it from the lower left-hand corner of detail ① to the left-hand side of detail ③ ?
26. What kind of setscrew is called for by detail ④ ?
27. Is the right-hand end of detail ② even, or flush, with the right-hand end of detail ① ?
28. What is the distance from the back edge of detail ② to the centerlines of the tapped holes?
29. How much tolerance is allowed for the location of the tapped holes?
30. Have any machined surfaces been called for on this part?

unit 18

PIN FASTENERS

Machine pins are widely used to hold or align assembled parts. The style of pin fastener selected depends upon the application required and the stress to which the pin may be subjected.

American National Standard dimensional information on machine pins is available from the American Society of Mechanical Engineers (ASME). The standard types of pin fasteners include taper pins, dowel pins, straight pins, grooved pins, spring pins, cotter pins, and clevis pins.

TAPER PINS

Taper pins are commonly used for easy alignment of parts to be assembled. A taper pin may be used in place of a key to align pulleys or gears on shafts where light loads are required.

The holes for taper pins must be drilled to a diameter slightly under the smaller diameter of the pin to be used, Figure 18-1A. The hole should be .002 to .005 undersize. The drilled hole is then tapered with a tapered reamer. The relationship between the drilled hole and the hole cut by the tapered reamer is shown at B in Figure 18-1. The finished hole is shown at C.

Taper pins have a standard taper of .250 ± .006 inch per foot and vary in size from a #7/0 pin to a #10 precision pin. Commercial nonprecision pins are available up to a #14 size, Table 18-1. Figure 18-2 illustrates an application of a standard taper pin.

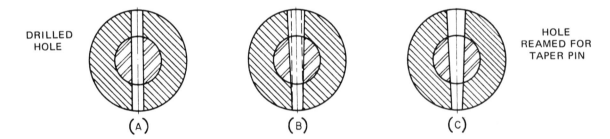

Fig. 18-1 Drilling and reaming for a taper pin

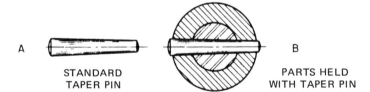

Fig. 18-2 Application of a straight taper pin

Table 18-1 Dimensions of Taper Pins (ANSI B18.8.2)

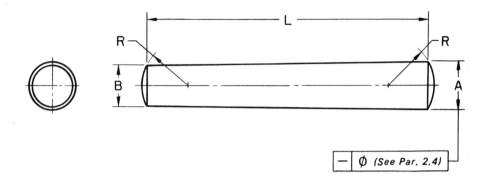

Pin Size Number and Basic Pin Diameter[1]	A Major Diameter (Large End)				R End Crown Radius	
	Commercial Class		Precision Class			
	Max	Min	Max	Min	Max	Min
7/0 0.0625	0.0638	0.0618	0.0635	0.0625	0.072	0.052
6/0 0.0780	0.0793	0.0773	0.0790	0.0780	0.088	0.068
5/0 0.0940	0.0953	0.0933	0.0950	0.0940	0.104	0.084
4/0 0.1090	0.1103	0.1083	0.1100	0.1090	0.119	0.099
3/0 0.1250	0.1263	0.1243	0.1260	0.1250	0.135	0.115
2/0 0.1410	0.1423	0.1403	0.1420	0.1410	0.151	0.131
0 0.1560	0.1573	0.1553	0.1570	0.1560	0.166	0.146
1 0.1720	0.1733	0.1713	0.1730	0.1720	0.182	0.162
2 0.1930	0.1943	0.1923	0.1940	0.1930	0.203	0.183
3 0.2190	0.2203	0.2183	0.2200	0.2190	0.229	0.209
4 0.2500	0.2513	0.2493	0.2510	0.2500	0.260	0.240
5 0.2890	0.2903	0.2883	0.2900	0.2890	0.299	0.279
6 0.3410	0.3423	0.3403	0.3420	0.3410	0.351	0.331
7 0.4090	0.4103	0.4083	0.4100	0.4090	0.419	0.399
8 0.4920	0.4933	0.4913	0.4930	0.4920	0.502	0.482
9 0.5910	0.5923	0.5903	0.5920	0.5910	0.601	0.581
10 0.7060	0.7073	0.7053	0.7070	0.7060	0.716	0.696
11 0.8600	0.8613	0.8593	*	*	0.870	0.850
12 1.0320	1.0333	1.0313	*	*	1.042	1.022
13 1.2410	1.2423	1.2403	*	*	1.251	1.231
14 1.5210	1.5223	1.5203	*	*	1.531	1.511

[1] Where specifying nominal pin size in decimals, zeros preceding decimal and in the fourth decimal place shall be omitted.
*Precision Class pins are not produced in these sizes.

Reference ANSI Y14.5 Dimensioning & Tolerancing.

Characteristics	Symbol
Straightness	—
Diameter	ϕ

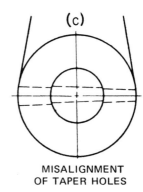

Fig. 18-3 Result of drilling and reaming parts separately

The alignment of mating parts is often critical to the assembly. Therefore, a print may call for parts to be drilled and reamed in assembly. The practice of using these notations will insure the alignment of both holes so that the pin will fit. Figure 18-3 shows the result of drilling and reaming parts separately.

DOWEL PINS

Dowels are precision-ground cylindrical pins which are used for alignment or guiding parts into position. Dowel pins are used when parts held with screws or bolts must be accurately fitted together, Figure 18-4.

Parts which must be held in assembly are most frequently drilled and reamed together. The two pieces are generally clamped in position, drilled, and reamed.

Dowel pins are available in standard diameters and lengths, Table 18-2. The pins are precision ground to a diameter which is .0002 inch over the nominal size. The dowels may be hardened or left soft.

STRAIGHT PINS

Straight pins are very similar to dowel pins except that they are not hardened and ground. Straight pins are made from cold drawn steel wire or rod. However, stainless steel, brass, or other metals may also be used when required.

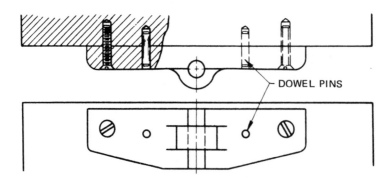

Fig. 18-4 Aligning parts with dowel pins

Table 18-2 Dimensions of Hardened Ground Machine Dowel Pins (ANSI B18.8.2)

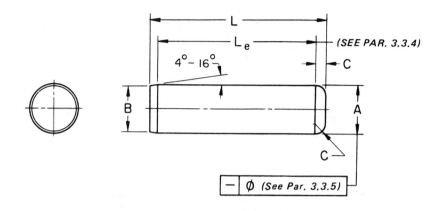

Nominal Size[1] or Nominal Pin Diameter		A Pin Diameter						B Point Diameter		C Crown Height or Radius		Double Shear Load Min, lb
		Standard Series Pins			Oversize Series Pins							Material
		Basic	Max	Min	Basic	Max	Min	Max	Min	Max	Min	Carbon or Alloy Steel
1/16	0.0625	0.0627	0.0628	0.0626	0.0635	0.0636	0.0634	0.058	0.048	0.020	0.008	800
*5/64	0.0781	0.0783	0.0784	0.0782	0.0791	0.0792	0.0790	0.074	0.064	0.026	0.010	1,240
3/32	0.0938	0.0940	0.0941	0.0939	0.0948	0.0949	0.0947	0.089	0.079	0.031	0.012	1,800
1/8	0.1250	0.1252	0.1253	0.1251	0.1260	0.1261	0.1259	0.120	0.110	0.041	0.016	3,200
*5/32	0.1562	0.1564	0.1565	0.1563	0.1572	0.1573	0.1571	0.150	0.140	0.052	0.020	5,000
3/16	0.1875	0.1877	0.1878	0.1876	0.1885	0.1886	0.1884	0.180	0.170	0.062	0.023	7,200
1/4	0.2500	0.2502	0.2503	0.2501	0.2510	0.2511	0.2509	0.240	0.230	0.083	0.031	12,800
5/16	0.3125	0.3127	0.3128	0.3126	0.3135	0.3136	0.3134	0.302	0.290	0.104	0.039	20,000
3/8	0.3750	0.3752	0.3753	0.3751	0.3760	0.3761	0.3759	0.365	0.350	0.125	0.047	28,700
7/16	0.4375	0.4377	0.4378	0.4376	0.4385	0.4386	0.4384	0.424	0.409	0.146	0.055	39,100
1/2	0.5000	0.5002	0.5003	0.5001	0.5010	0.5011	0.5009	0.486	0.471	0.167	0.063	51,000
5/8	0.6250	0.6252	0.6253	0.6251	0.6260	0.6261	0.6259	0.611	0.595	0.208	0.078	79,800
3/4	0.7500	0.7502	0.7503	0.7501	0.7510	0.7511	0.7509	0.735	0.715	0.250	0.094	114,000
7/8	0.8750	0.8752	0.8753	0.8751	0.8760	0.8761	0.8759	0.860	0.840	0.293	0.109	156,000
1	1.0000	1.0002	1.0003	1.0001	1.0010	1.0011	1.0009	0.980	0.960	0.333	0.125	204,000

[1] Where specifying nominal size as basic diameter, zeros preceding decimal and in the fourth decimal place shall be omitted.

*Nonpreferred sizes, not recommended for use in new designs.

Reference ANSI Y14.5 Dimensioning & Tolerancing.

Characteristics	Symbols
Straightness	—
Diameter	ϕ

The diameter of straight pins conforms to the diameter of the wire or rod from which they are made, Table 18-3.

GROOVED PINS

Grooved pins have longitudinal grooves which are pressed or rolled into the diameter. The operation which forms the groove expands the diameter of the pin by displacing material to specified limits. American National Standard groove pins have three grooves of equal depth, length, and shape equally spaced around the diameter.

Table 18-3 Dimensions of Hardened Ground Production Dowel Pins (ANSI B18.8.2)

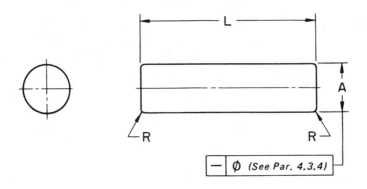

Nominal Size[1] or Nominal Pin Diameter		A Pin Diameter			R Corner Radius		Double Shear Load Min, Lb Material Carbon Steel
		Basic	Max	Min	Max	Min	
1/16	0.0625	0.0627	0.0628	0.0626	0.020	0.010	790
3/32	0.0938	0.0939	0.0940	0.0938	0.020	0.010	1,400
7/64	0.1094	0.1095	0.1096	0.1094	0.020	0.010	1,900
1/8	0.1250	0.1252	0.1253	0.1251	0.020	0.010	2,600
5/32	0.1562	0.1564	0.1565	0.1563	0.020	0.010	4,100
3/16	0.1875	0.1877	0.1878	0.1876	0.020	0.010	5,900
7/32	0.2188	0.2189	0.2190	0.2188	0.020	0.010	7,600
1/4	0.2500	0.2502	0.2503	0.2501	0.020	0.010	10,000
5/16	0.3125	0.3127	0.3128	0.3126	0.020	0.010	16,000
3/8	0.3750	0.3752	0.3753	0.3751	0.020	0.010	23,000

[1] Where specifying nominal pin size in decimals, zeros preceding decimal and in the fourth decimal place shall be omitted.

Reference ANSI Y14.5 Dimensioning & Tolerance.

Characteristics   Symbol
Straightness   —
Diameter ϕ

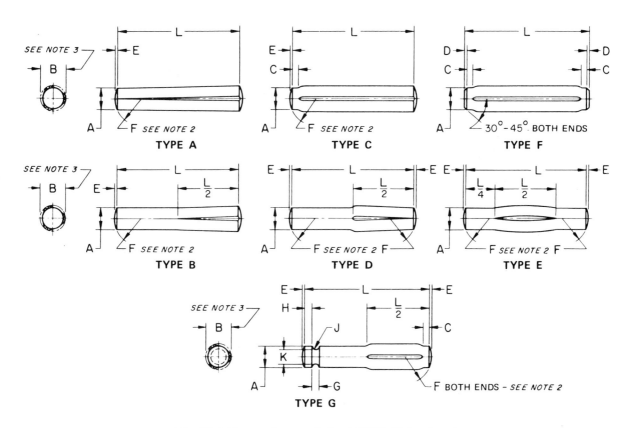

Fig. 18-5 Types of grooved pins (ANSI B18.8.2-1978)

Grooved pins provide a locking fit when forced into a drilled hole. The raised ridge on the pin provides the proper fit between the pin and hole. Holes do not need to be reamed when grooved pins are used. A variety of styles of grooved pins is available for various operations, Figure 18-5. Dimensional data is provided in Table 18-4.

CLEVIS PINS

Clevis pins are straight diameter pins with a head at one end. The diameter of the pin is slightly undersize so that the pin may slip freely through a standard diameter hole. For example, a 3/8 (.375) diameter pin has a body size of .373. Clevis pins have cotter pin holes drilled at the end opposite the head. The cotter pin is used to hold the clevis pin in place, Figure 18-6.

SPRING PINS

Spring pins may be slotted or coiled. This type of spring is used in drilled holes and reaming is not necessary. The spring pins are not truly round in the free state due to the manufactured slot or overlapping coils. Spring pins have the ability to compress and expand against the sides of the drilled hole. This provides a tight fit regardless of slight variations in drilled hole size. Tables 18-5 and 18-6 provide descriptive information regarding spring pins.

Table 18-4 Dimensions of Grooved Pins (ANSI B18.8.2)

Nominal Size[1] or Basic Pin Diameter	A Pin Diameter		C Pilot Length	D[2] Chamfer Length	E[2] Crown Height		F[2] Crown Radius		G Neck Width		H Shoulder Length		J Neck Radius	K Neck Diameter	
	Max	Min	Ref	Min	Max	Min	Max	Min	Max	Min	Max	Min	Ref	Max	Min
1/32 0.0312	0.0312	0.0302	0.015	...	...	...	...	...	...	...	...	...	...	...	...
3/64 0.0469	0.0469	0.0459	0.031	...	...	...	...	...	...	...	...	...	...	...	...
1/16 0.0625	0.0625	0.0615	0.031	0.016	0.0115	0.0015	0.088	0.068	...	...	...	...	...	...	...
5/64 0.0781	0.0781	0.0071	0.031	0.016	0.0137	0.0037	0.104	0.084	...	...	...	...	...	...	...
3/32 0.0938	0.0938	0.0928	0.031	0.016	0.0141	0.0041	0.135	0.115	0.038	0.028	0.041	0.031	0.016	0.067	0.057
7/64 0.1094	0.1094	0.1074	0.031	0.016	0.0160	0.0060	0.150	0.130	0.038	0.028	0.041	0.031	0.016	0.082	0.072
1/8 0.1250	0.1250	0.1230	0.031	0.016	0.0180	0.0080	0.166	0.146	0.069	0.059	0.041	0.031	0.031	0.088	0.078
5/32 0.1563	0.1563	0.1543	0.062	0.031	0.0220	0.0120	0.198	0.178	0.069	0.059	0.057	0.047	0.031	0.109	0.099
3/16 0.1875	0.1875	0.1855	0.062	0.031	0.0230	0.0130	0.260	0.240	0.069	0.059	0.057	0.047	0.031	0.130	0.120
7/32 0.2188	0.2188	0.2168	0.062	0.031	0.0270	0.0170	0.291	0.271	0.101	0.091	0.072	0.062	0.047	0.151	0.141
1/4 0.2500	0.2500	0.2480	0.062	0.031	0.0310	0.0210	0.322	0.302	0.101	0.091	0.072	0.062	0.047	0.172	0.162
5/16 0.3125	0.3125	0.3105	0.094	0.047	0.0390	0.0290	0.385	0.365	0.132	0.122	0.104	0.094	0.062	0.214	0.204
3/8 0.3750	0.3750	0.3730	0.094	0.047	0.0440	0.0340	0.479	0.459	0.132	0.122	0.135	0.125	0.062	0.255	0.245
7/16 0.4375	0.4375	0.4355	0.094	0.047	0.0520	0.0420	0.541	0.521	0.195	0.185	0.135	0.125	0.094	0.298	0.288
1/2 0.5000	0.5000	0.4980	0.094	0.047	0.0570	0.0470	0.635	0.615	0.195	0.185	0.135	0.125	0.094	0.317	0.307

[1] Where specifying nominal size in decimals, zeros preceding decimal and in the fourth decimal place shall be omitted.
[2] Pins in 1/32 and 3/64 in. sizes of any length and all sizes 1/4 in. nominal length, or shorter, are not crowned or chamfered. See Paragraph 7.4 of General Data. Alloy steel pins of all types shall have chamfered ends conforming with Type F pins, included within the pin length.

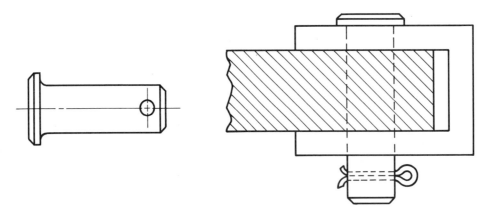

Fig. 18-6 Clevis pins

Table 18-5 Dimensions of Slotted Type Spring Pins (ANSI B18.8.2)

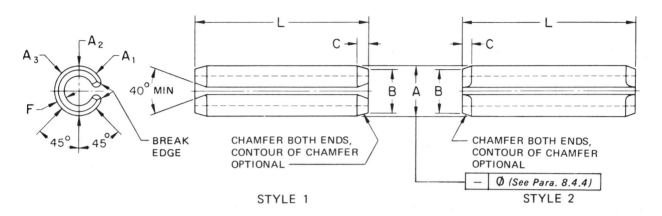

OPTIONAL CONSTRUCTIONS

Nominal Size[1] or Basic Pin Diameter		A Pin Diameter		B Chamfer Diameter	C Chamfer Length		F Stock Thickness	Recommended Hole Size		Double Shear Load, Min, lb		
										Material		
		Max[2]	Min[3]	Max	Max	Min	Basic	Max	Min	AISI 1070-1095[4] and AISI 420	AISI 302	Beryllium Copper
1/16	0.062	0.069	0.066	0.059	0.028	0.007	0.012	0.065	0.062	425	350	270
5/64	0.078	0.086	0.083	0.075	0.032	0.008	0.018	0.081	0.078	650	550	400
3/32	0.094	0.103	0.099	0.091	0.038	0.008	0.022	0.097	0.094	1,000	800	660
1/8	0.125	0.135	0.131	0.122	0.044	0.008	0.028	0.129	0.125	2,100	1,500	1,200
9/64	0.141	0.149	0.145	0.137	0.044	0.008	0.028	0.144	0.140	2,200	1,600	1,400
5/32	0.156	0.167	0.162	0.151	0.048	0.010	0.032	0.160	0.156	3,000	2,000	1,800
3/16	0.188	0.199	0.194	0.182	0.055	0.011	0.040	0.192	0.187	4,400	2,800	2,600
7/32	0.219	0.232	0.226	0.214	0.065	0.011	0.048	0.224	0.219	5,700	3,550	3,700
1/4	0.250	0.264	0.258	0.245	0.065	0.012	0.048	0.256	0.250	7,700	4,600	4,500
5/16	0.312	0.328	0.321	0.306	0.080	0.014	0.062	0.318	0.312	11,500	7,095	6,800
3/8	0.375	0.392	0.385	0.368	0.095	0.016	0.077	0.382	0.375	17,600	10,000	10,100
7/16	0.438	0.456	0.448	0.430	0.095	0.017	0.077	0.445	0.437	20,000	12,000	12,200
1/2	0.500	0.521	0.513	0.485	0.110	0.025	0.094	0.510	0.500	25,800	15,500	16,800
5/8	0.625	0.650	0.640	0.608	0.125	0.030	0.125	0.636	0.625	46,000[4]	18,800	...
3/4	0.750	0.780	0.769	0.730	0.150	0.030	0.150	0.764	0.750	66,000[4]	23,200	...

[1] Where specifying nominal size in decimals, zeros preceding the decimal shall be omitted.
[2] Maximum diameter shall be checked by GO ring gage.
[3] Minimum diameter shall be average of three diameters measured at points illustrated. A min = $\frac{A_1 + A_2 + A_3}{3}$
[4] Sizes 5/8 in. (0.625) and larger are produced from AISI 6150H alloy steel, not AISI 1070-1095.

Characteristics	Symbol
Straightness	—
Diameter	ϕ

Table 18-6 Dimensions of Coiled Type Spring Pins (ANSI B18.8.2)

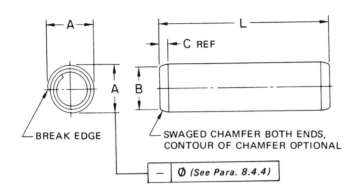

Nominal Size[1] or Basic Pin Diameter		A Pin Diameter						B Chamfer		C		Recommended Hole Size		Double Shear Load, Min, lb					
														Standard Duty		Heavy Duty		Light Duty	
														Material					
		Standard Duty		Heavy Duty		Light Duty		Dia	Length	Max	Min			AISI 1070-1095[4,5] and AISI 420	AISI 302	AISI 1070-1095[5] and AISI 420	AISI 302	AISI 1070-1095 and AISI 420	AISI 302
		Max[2]	Min[3]	Max[2]	Min[3]	Max[2]	Min[3]	Max	Ref	Max	Min								
1/32	0.031	0.035	0.033	...	...	...	...	0.029	0.024	0.032	0.031			75[4]	60	...	...	...	...
	0.039	0.044	0.041	...	...	...	...	0.037	0.024	0.040	0.039			120[4]	100	...	...	...	...
3/64	0.047	0.052	0.049	...	...	...	...	0.045	0.024	0.048	0.046			170[4]	140	...	...	...	...
	0.052	0.057	0.054	...	...	...	...	0.050	0.024	0.053	0.051			230[4]	190	...	...	...	...
1/16	0.062	0.072	0.067	0.070	0.066	0.073	0.067	0.059	0.028	0.065	0.061			300	250	450	350	...	135
5/64	0.078	0.088	0.083	0.086	0.082	0.089	0.083	0.075	0.032	0.081	0.077			475	400	700	550	...	225
3/32	0.094	0.105	0.099	0.103	0.098	0.106	0.099	0.091	0.038	0.097	0.093			700	550	1,000	800	375	300
7/64	0.109	0.120	0.114	0.118	0.113	0.121	0.114	0.106	0.038	0.112	0.108			950	750	1,400	1,125	525	425
1/8	0.125	0.138	0.131	0.136	0.130	0.139	0.131	0.121	0.044	0.129	0.124			1,250	1,000	2,100	1,700	675	550
5/32	0.156	0.171	0.163	0.168	0.161	0.172	0.163	0.152	0.048	0.160	0.155			1,925	1,550	3,000	2,400	1,100	875
3/16	0.188	0.205	0.196	0.202	0.194	0.207	0.196	0.182	0.055	0.192	0.185			2,800	2,250	4,400	3,500	1,500	1,200
7/32	0.219	0.238	0.228	0.235	0.226	0.240	0.228	0.214	0.065	0.224	0.217			3,800	3,000	5,700	4,600	2,100	1,700
1/4	0.250	0.271	0.260	0.268	0.258	0.273	0.260	0.243	0.065	0.256	0.247			5,000	4,000	7,700	6,200	2,700	2,200
5/16	0.312	0.337	0.324	0.334	0.322	0.339	0.324	0.304	0.080	0.319	0.308			7,700	6,200	11,500	9,200	4,440	3,500
3/8	0.375	0.403	0.388	0.400	0.386	0.405	0.388	0.366	0.095	0.383	0.370			11,200	9,000	17,600	14,000	6,000	5,000
7/16	0.438	0.469	0.452	0.466	0.450	0.471	0.452	0.427	0.095	0.446	0.431			15,200	13,000	22,500	18,000	8,400	6,700
1/2	0.500	0.535	0.516	0.532	0.514	0.537	0.516	0.488	0.110	0.510	0.493			20,000	16,000	30,000	24,000	11,000	8,800
5/8	0.625	0.661	0.642	0.658	0.640	...	...	0.613	0.125	0.635	0.618			31,000[5]	25,000	46,000[5]	37,000	...	...
3/4	0.750	0.787	0.768	0.784	0.766	...	...	0.738	0.150	0.760	0.743			45,000[5]	36,000	66,000[5]	53,000	...	...

[1] Where specifying nominal size in decimals, zeros preceding decimal shall be omitted.
[2] Maximum diameter shall be checked by GO ring gage.
[3] Minimum diameter shall be checked by NOT GO ring gage.
[4] Sizes 1/32 in. (0.031) through 0.052 in. are not available in AISI 1070–1095 carbon steel.
[5] Sizes 5/8 in. (0.625) and larger are produced from AISI 6150 alloy steel, not AISI 1070–1095 carbon steel.

Characteristics	Symbol
Straightness	—
Diameter	ϕ

COTTER PINS

Cotter pins are classified as mechanical accessories. They are commonly used as pin fasteners in the assembly of machine parts where great accuracy is not required.

Cotter pins may be represented on drawings in many different ways as shown in Figure 18-7A, B, and C.

The commercial sizes of cotter pins are specified by wire gauge number. The ANSI standard sizes are given in fractional inch dimensions.

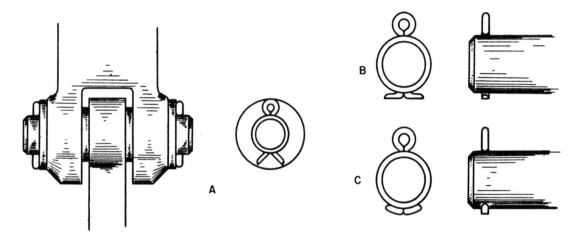

Fig. 18-7 Representing cotter pins in assembly

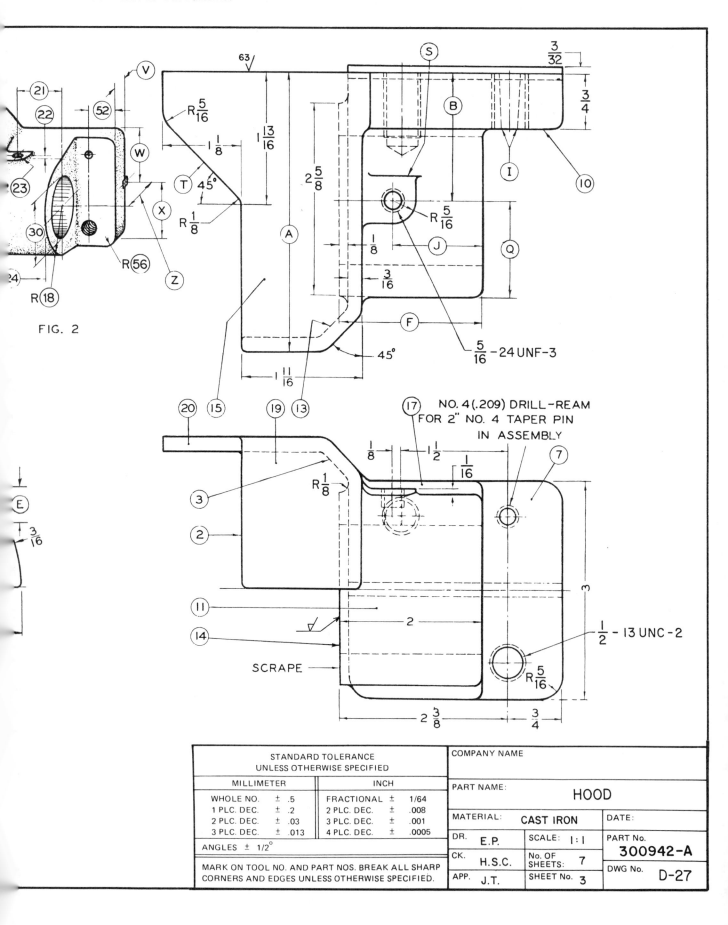

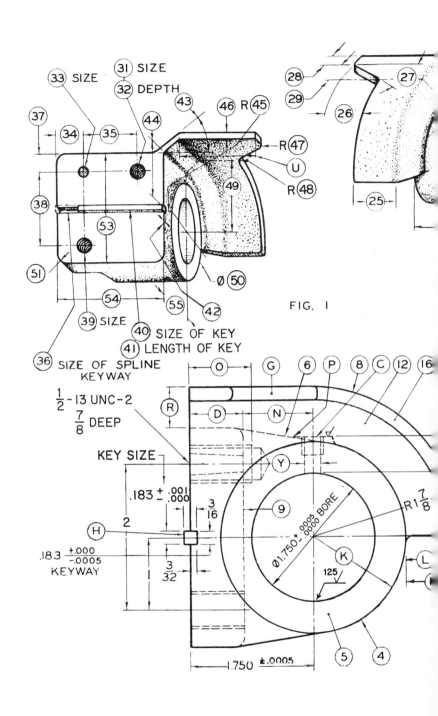

FIG. 1

ASSIGNMENT D-27: HOOD

1. What views are shown?

2. What is to fit into tapped hole Ⓒ ?
3. Which is surface ③ in the left-side view?
4. Locate surface ② in the left-side view.
5. What surface in the front view is ④ ?
6. What surface in the left-side view is ⑭ ?
7. What surface in the top view and the front view is ⑨ ?
8. What surface in the top view and the front view is ⑧ ?
9. What surface in the top view is ⑫ ?
10. What is dimension Ⓐ ?
11. What is dimension Ⓑ ?
12. What is dimension Ⓕ ?
13. What is dimension Ⓓ ?
14. What is dimension Ⓔ ?
15. What is the name of part Ⓗ ?
16. What is its purpose?
17. What do the dotted lines at Ⓘ represent?
18. What is dimension Ⓙ ?
19. What surface in the front view is line ⑥ ?
20. What radius is Ⓚ ?
21. What distance is Ⓛ ?
22. What distance is Ⓜ ?
23. What distance is Ⓝ ?
24. What is the depth of the tapped hole at Ⓞ ?
25. What line in the top view indicates point Ⓟ ?
26. What is dimension Ⓠ ?
27. What is dimension Ⓡ ?
28. What edges or surfaces in the front view and the left-side view does line Ⓣ represent?
29. What is the diameter of tap drill Ⓨ ?
30. Determine dimensions or operations at

    Ⓤ ㉗ ㊲ ㊼   Ⓤ = _____   ㉙ = _____   ㊸ = _____
    Ⓥ ㉘ ㊳ ㊽   Ⓥ = _____   ㉚ = _____   ㊹ = _____
    Ⓦ ㉙ ㊴ ㊾   Ⓦ = _____   ㉛ = _____   ㊺ = _____
    Ⓧ ㉚ ㊵ ㊿   Ⓧ = _____   ㉜ = _____   ㊻ = _____
    Ⓩ ㉛ ㊶ ㊱   Ⓩ = _____   ㉝ = _____   ㊼ = _____
    ⑱ ㉜ ㊷ ㊵   ⑱ = _____   ㉞ = _____   ㊽ = _____
    ㉑ ㉝ ㊸ ㊳   ㉑ = _____   ㉟ = _____   ㊾ = _____
    ㉒ ㉞ ㊹ ㊴   ㉒ = _____   ㊱ = _____   ㊿ = _____
    ㉓ ㉟ ㊺ ㊵   ㉓ = _____   ㊲ = _____   ㊱ = _____
    ㉔ ㊱ ㊻ ㊶   ㉔ = _____   ㊳ = _____   ㊲ = _____
    ㉕   ㉕ = _____   ㊴ = _____   ㊳ = _____
    ㉖   ㉖ = _____   ㊵ = _____   ㊴ = _____
                        ㉗ = _____   ㊶ = _____   ㊵ = _____
                        ㉘ = _____   ㊷ = _____   ㊶ = _____

unit 19

SPRINGS

A spring is a mechanical device commonly used in machine design and construction. Springs provide a source of energy which may be used to resist extension, compression, or twisting of parts. Materials used in spring manufacture must be able to withstand the fatigue and stress created by frequent distortion. High carbon spring steel, alloy spring steel, and stainless spring steel are among the most common materials used.

HELICAL SPRINGS

Springs are classified in two basic categories: coil or helical springs and flat springs. Helical springs may be of the following types: compression, extension, or torsion. Each spring type is used for applications where resistance to movement is required.

Compression springs resist part compression. Extension springs resist the pulling apart of objects. Torsion springs resist the twisting forces or torque applied to an object. Helical springs may be cylindrical or conical in shape or a combination of both. Figure 19-1 illustrates some of the common types of helical springs.

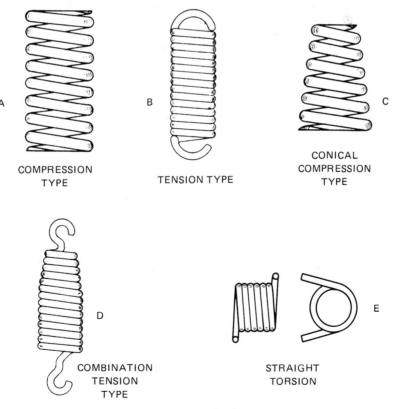

Fig. 19-1 Coil or helical springs

160 Unit 19 Springs

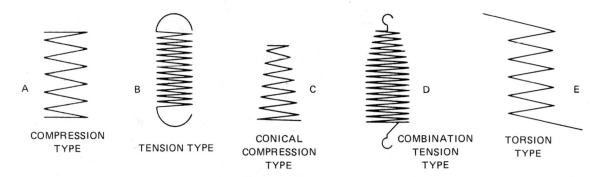

Fig. 19-2 Simplified method of representing helical springs

The true projection of a helical spring is usually not drawn because of the time involved in making the drawing. Instead, schematic or straight line drawings are used to represent the various spring types. Straight line drawings simplify the representation while still providing all necessary information. Figure 19-2 illustrates typical schematic representations of the various spring types.

FLAT SPRINGS

Flat springs are made from flat spring steel material and often are specifically designed for the job application, Figure 19-3. Commercially available flat springs such as spring washers or leaf-type springs are incorporated in designs where possible.

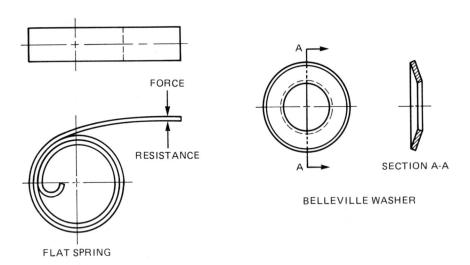

Fig. 19-3 Flat springs

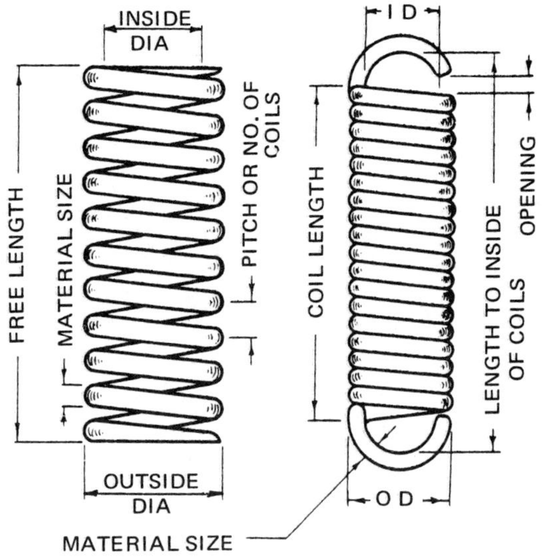

Fig. 19-4 Information given on drawings for springs

SPECIFICATION OF SPRINGS ON DRAWINGS

When springs are required as part of a machine assembly, the following information must be given on the print, Figure 19-4:

- Size, shape, and spring material
- Spring diameter (either inside or outside)
- Pitch or number of coils
- Shape of the ends
- Length

For example, a typical drawing callout may read as follows:

> REQUIRED: One helical tension spring 3" long (or number of coils), inside diameter 1/2", pitch 3/16", 18 B & S GA, spring brass wire.

The *pitch* of a coil spring is the distance from the center of one coil to the center of the next. The sizes of spring wires are designated by gauge numbers and also in decimal parts of an inch. Tables of sizes can be found in handbooks.

Springs are made to a dimension of either outside diameter (if the spring works in a hole) or inside diameter (if the spring works on a rod). In some cases, the mean diameter is specified for computation purposes.

162 Unit 19 Springs

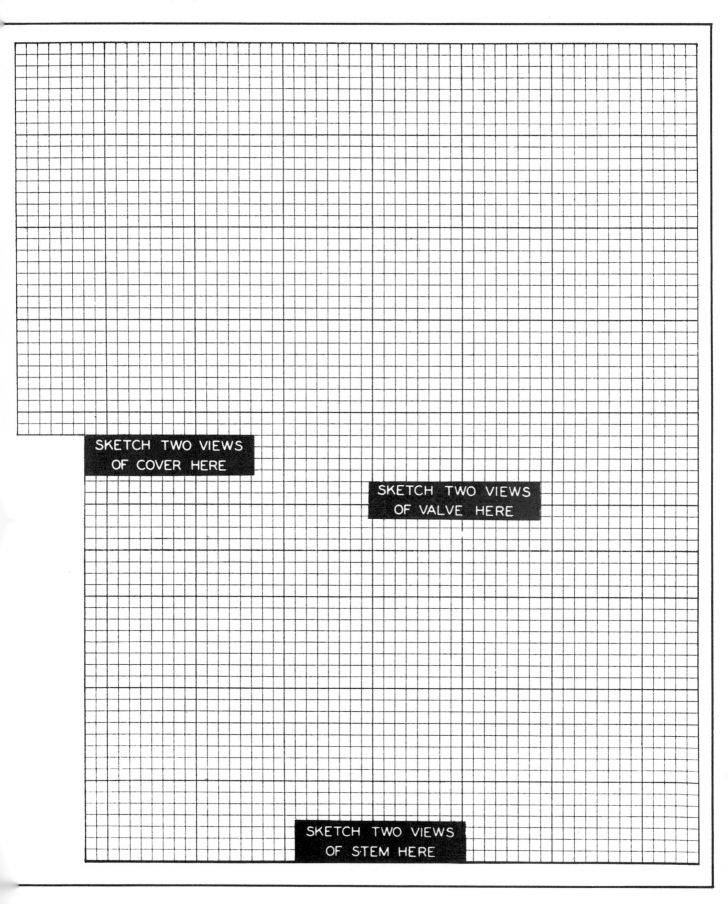

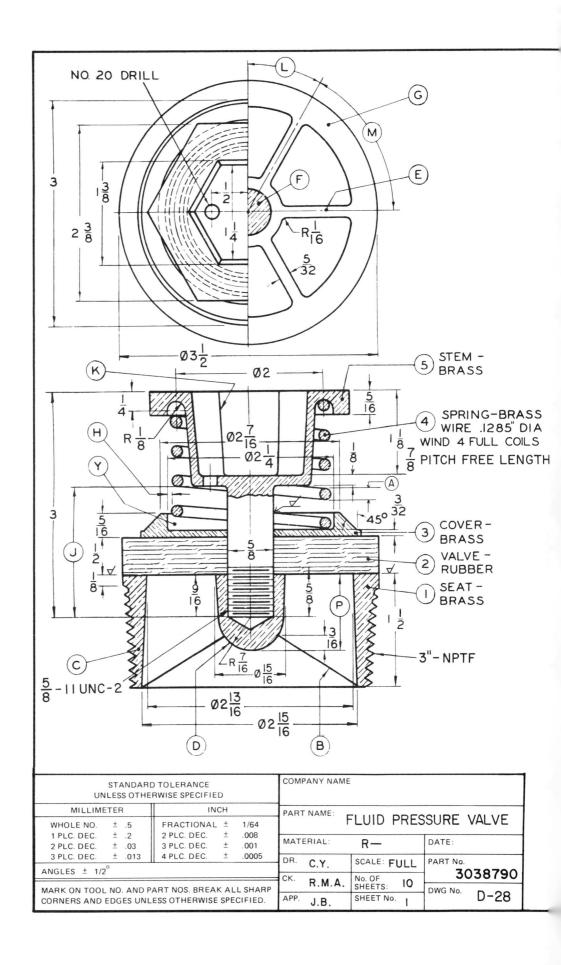

ASSIGNMENT D-28: FLUID PRESSURE VALVE

1. Sketch two views of each of the parts asked for in the space provided on the drawing of the Fluid Pressure Valve.
2. How many separate parts are shown on the valve assembly?
3. What material is specified for part ①?
4. What material is specified for part ②?
5. What material is specified for part ③?
6. What material is specified for part ④?
7. What material is part ⑤ made of?
8. What is the length of the spring when the Valve is closed?
9. Determine distance Ⓐ.
10. What is the overall free length of the spring?
11. Locate Ⓔ in the front view.
12. How many supporting ribs are there connecting Ⓓ to Ⓒ?
13. How thick are these ribs?
14. Locate parts Ⓕ and Ⓖ in the front view.
15. How many threads per inch are there on the stem, part ⑤? How are these threads designated? State the size and the kind.
16. What is the nominal size of the pipe thread?
17. Give the length of the pipe thread.
18. Determine the clearance distance Ⓗ as shown.
19. Determine distance Ⓙ.
20. Give the approximate diameter at the small end of the pipe thread.
21. Indicate line Ⓚ by the circled letter Ⓚ in each view of the freehand sketch of the stem.
22. Indicate surface Ⓨ by the circled letter Ⓨ in each view of the freehand sketch of the cover.
23. Determine distance Ⓟ.
24. What is angle Ⓛ?
25. What is angle Ⓜ?

unit 20

SWIVELS, UNIVERSAL JOINTS, AND BEARINGS

SWIVELS

A swivel is composed of two or more pieces which are so constructed that either part rotates in relation to the other about a common axis, Figure 20-1.

UNIVERSAL JOINT

A universal joint, Figure 20-2, is composed of three or more pieces which are designed to permit the free rotation of two shafts whose axes deviate from a straight line.

In practice, universal joints are constructed in many different forms, the design of which is influenced by the requirements of the working mechanism and the cost of the parts.

Universal joints are used by the designer where a rotating or swinging motion is desired or where power must be delivered along shafts which are not in a straight line.

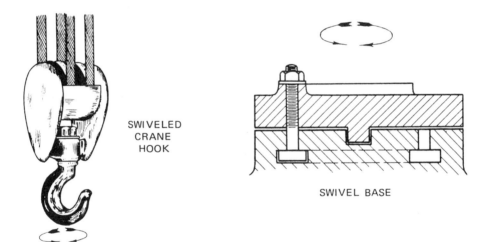

Fig. 20-1 Types of swivels

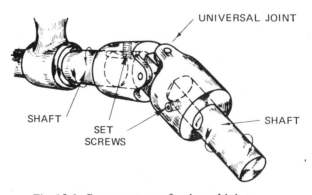

Fig. 20-2 Common type of universal joint

Unit 20 Swivels, Universal Joints, and Bearings 165

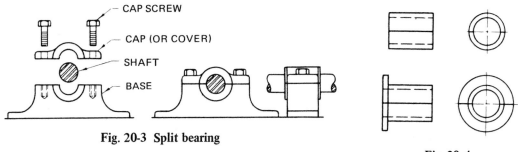

Fig. 20-3 Split bearing

Fig. 20-4
Two types of split bushings

BORING SPLIT BEARINGS

In the design and construction of bearings, it is sometimes necessary to split the bearing to facilitate assembling, to permit adjusting, and also to make possible the replacing of worn parts. The split type of bearing allows the shaft or spindle to be set in one half of the bearing while the other half, or cover, is later secured in position, Figure 20-3.

If the bearing, Figure 20-3, or the bushing shown in Figure 20-4 are to be made from two parts, it is necessary that they be fastened together before the hole is bored or reamed. If this were not done, the machining operation would be more difficult and the two parts might not make a perfectly round bearing when assembled, Figure 20-5.

One method used to give longer life to the bearing is to insert very thin strips of metal between the base and cover halves before boring. These thin strips, which vary in thickness, are called *shims*.

When a bearing is shimmed, the same number of pieces of corresponding thickness are used on both sides of the bearing.

As the hole wears, one or more pairs of these shims may be removed to compensate for the wear, Figure 20-6.

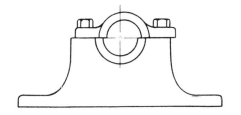

Fig. 20-5 Bearing halves incorrectly matched

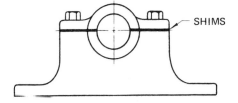

Fig. 20-6 Shimmed bearing

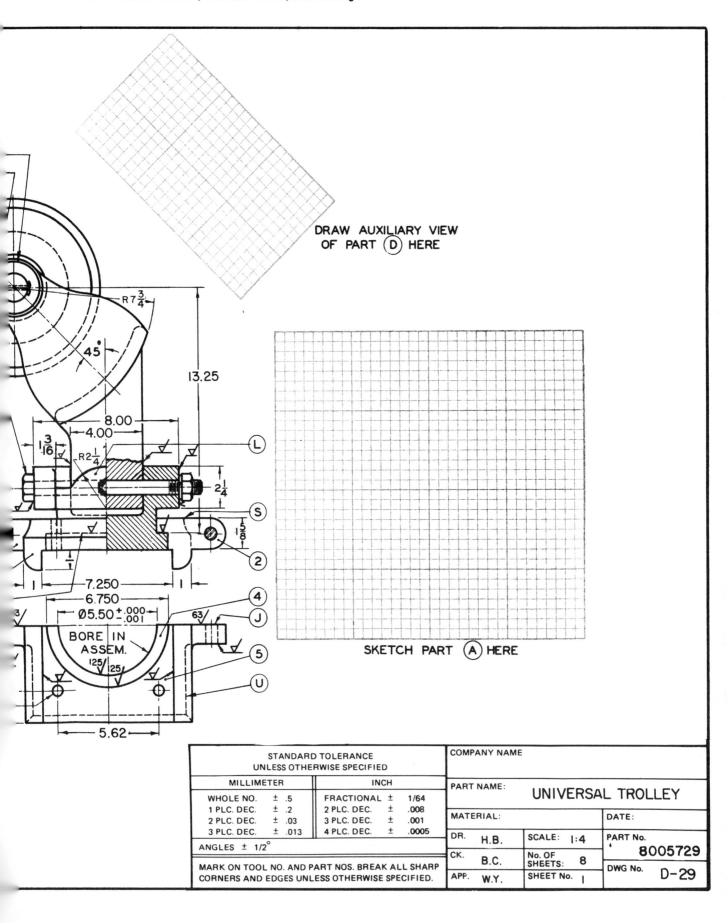

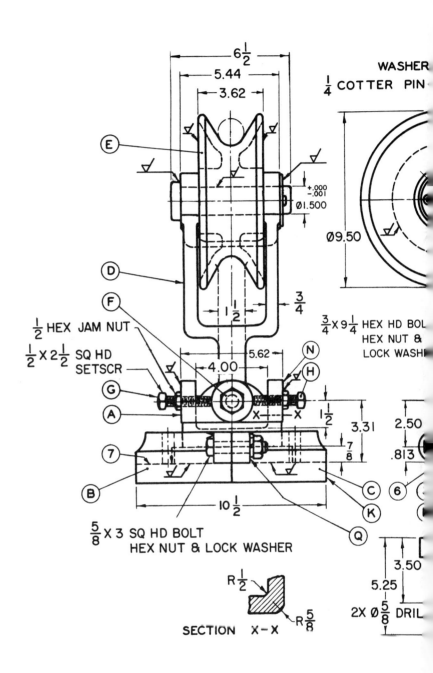

ASSIGNMENT D-29: UNIVERSAL TROLLEY

1. Draw a freehand auxiliary view of part Ⓓ in the space provided on the drawing of the Universal Trolley.
2. Make a freehand sketch, with three views, of part Ⓐ in the space provided on the Universal Trolley drawing. Do not dimension.
3. What line or surface in the bottom view represents surface ②?   _____
4. Is surface ③ shown in the bottom view?   _____
5. What line or surface in the left-side view represents surface ③?   _____
6. How many different parts are used to make the Trolley?   _____
7. What line or surface in the left-side view represents surface ⑥?   _____
8. How many tapped holes are there in all of the parts of the trolley? (Do not include the nuts.)   _____
9. Give the total number of drilled and bored holes in the Trolley, not including the tapped holes or nuts and washers.   _____
10. What line or surface in the bottom view represents point Ⓢ?   _____
11. How many actual parts in the Trolley assembly are standard mechanical accessories?   _____
12. What line or surface in the front view represents surface ④?   _____
13. What line or surface in the left-side view represents surface ⑤?   _____
14. What is the overall height of the Trolley?   _____
15. What is the outside diameter of the lockwasher for the largest bolt?   _____
16. What line or surface in the left-side view represents surface Ⓛ?   _____
17. What specifications are required for the cotter pin?   _____
18. Determine the outside diameter of Ⓔ.   _____
19. Why is the $5.500 \begin{smallmatrix} +.001 \\ -.000 \end{smallmatrix}$ dimension bored in assembly?   _____
20. What thread representation is used on the threaded fasteners?   _____

unit 21

WORM GEARING

The interpretation of worm gear drawings requires the following skills:

1. An understanding of the methods by which certain features of worms and worm gears are represented.

2. A knowledge of both gear tooth parts and worm threads.

3. The ability to use the mathematical rules for computing the data required in the construction of the worm gear and worm.

A *worm gear,* or worm wheel as it is sometimes called, is a cylindrical disk or wheel. This disk has a specific number of equally spaced and uniformly shaped teeth which are cut, at an angle to the axis of the gear, into an outer rim. The rim is concave so that the threads of the mating worm fit into the teeth on the gear.

A *worm* is a special form of screw thread in which the cross section resembles a rack gear tooth of corresponding pitch. The linear pitch of the worm corresponds to the pitch of any screw thread. All of the characteristics of the worm thread must be related to the teeth of the worm gear with which it meshes.

The worm and worm gear are used to transmit power between shafts when the shaft axes are nonintersecting and usually at right angles to each other. The principal advantage of worm gearing is that it is possible to obtain a large reduction in velocity. This velocity reduction occurs because in one revolution of a single-threaded worm, the worm wheel is advanced only one tooth.

REPRESENTATION OF WORM GEARS

Worm gears may be represented conventionally as shown on the Worm Gear drawing in the Assignment. The front view and the section view usually contain the dimensions which, when combined with the data contained in a table on the drawing, furnish the information necessary for the construction of the gear.

The chamfered, or beveled worm wheel rim shown in Figure 21-1A is used less frequently in modern machine tools as it has largely been replaced by the type of rim shown in Figure 21-1B. This rim is almost square, except for the corners which are rounded to a radius equal to one-fourth of the circular pitch. The rounded rim is better suited for power transmission because the tooth contact area is increased over that of the beveled design. Another advantage is that the rounded rim requires less machining than the beveled rim.

*American Gear Manufacturer's Association

Fig. 21-1 Comparison of beveled and modern-type worm wheel rim

WORM GEARING PARTS, SYMBOLS, AND TERMS

If a plane is passed through the center of the worm perpendicular to the axis of the worm wheel, the worm section will be similar to a rack. The worm wheel section will resemble a spur gear. Some of the terms and formulas which are used to compute the spur gear and rack dimensions are applicable to worm gearing and are given in Table 21-1. See Table 21-2 for worm gear rules and formulas.

Worm screws can be either single- or multiple-threaded types. Therefore, the machinist must be careful when using the terms *pitch* and *lead*. The pitch, or more correctly, the *linear pitch*, is the distance between a given point on one thread to a corresponding point on the next thread. This distance is measured parallel to the axis of the worm.

Lead is the distance which any one thread on the worm advances in one revolution. The pitch and lead of the worm are related as follows:

- In a single-threaded worm, the linear pitch and lead are equal.
- For a double-threaded worm, the lead is twice the linear pitch.
- In all other multiple threads, the lead is equal to the multiple times the linear pitch.

Table 21-1 Symbols and Terms Used in Worm Gearing Formulas

P =	Circular pitch of worm wheel and linear or axial pitch of worm	S =	Addendum (height of worm tooth above pitch line)
P' =	Normal pitch	H =	Throat diameter
L =	Lead of worm	G =	Throat radius
A =	Helix or lead angle of worm and gashing angle for gear	T =	Width of thread tool at small end
D =	Pitch diameter of worm gear	W =	Whole depth of thread or tooth
D' =	Pitch diameter of worm	R =	Root diameter of worm
N =	Number of teeth in worm gear	B =	Normal thickness of tooth
N' =	Number of threads in worm (Single or multiple)	F =	Face width
		E =	Corner radius
O =	Outside diameter of worm gear or wheel	X =	Minimum threaded length of worm
O' =	Outside diameter of worm	C =	Center distance

Table 21-2 Rules and Formulas for Worm Gearing

REQUIRED DIMENSION	GIVEN DATA	RULE	FORMULA
Linear or Axial Pitch of Worm or Circular Pitch of Gear	Lead of worm Number of threads (single or multiple)	Divide the lead by the number of threads.	$P = \dfrac{L}{N'}$
	Pitch diameter of gear Number of teeth in gear	Multiply pitch diameter by 3.1416 and divide by the number of teeth.	$P = \dfrac{3.1416\,D}{N}$
Normal Pitch	Linear pitch Helix angle	Multiply linear pitch by cosine of helix angle.	$P' = P(\cos A)$
Lead of Worm	Linear pitch Number of threads (single or multiple)	Multiply linear pitch by number of threads.	$L = N'P$
	Pitch diameter of worm Helix angle	Multiply pitch diameter by 3.1416 and then by the tangent of the helix angle.	$L = 3.1416\,D'\,(\tan A)$
Addendum of Worm or Gear	Linear pitch	Multiply linear pitch by .3183. Note: For helix angles less than 15°.	$S = .3183\,P$
	Normal pitch	Multiply normal pitch by .3183. Note: For helix angles greater than 15°.	$S = .3183\,P'$
Pitch Diameter of Worm	Addendum Outside diameter of worm	Subtract twice the addendum from the outside diameter.	$D' = O' - 2S$
Pitch Diameter of Worm Gear	Circular pitch Number of teeth in gear	Multiply circular pitch by number of teeth, then divide by 3.1416.	$D = \dfrac{NP}{3.1416}$
Throat Diameter	Addendum Pitch diameter of worm gear	Add twice the addendum to the pitch diameter.	$H = D + 2S$
	Circular or axial pitch Pitch diameter of worm gear	Multiply circular pitch by .572 and add to the pitch diameter. Note: Recommended for triple and Quadruple threads by A.G.M.A.	$H = D + (.572P)$
Outside Diameter of Worm Gear	Addendum Pitch diameter of gear	Add three times the addendum to the pitch diameter. Note: Recommended practice of David Brown & Sons Ltd.	$O = D + 3S$
	Throat diameter Circular pitch	Add to the throat diameter .4775 times the circular pitch. Note: Recommended for single and double threads by A.G.M.A.	$O = H + (.4775P)$
		Add to the throat diameter .3183 times the circular pitch. Note: Recommended for triple and quadruple threads by A.G.M.A.	$O = H + (.3183P)$
Outside Diameter of Worm	Addendum Pitch diameter of worm	Add twice the addendum to the pitch diameter.	$O' = D' + 2S$
Throat Radius	Pitch diameter of worm Addendum	Divide pitch diameter of worm by 2 then subtract the addendum.	$G = \dfrac{D'}{2} - S$
Helix or Lead Angle of Worm	Pitch diameter of worm Lead	The cotangent of the helix angle of the worm is equal to the product of 3.1416 times the pitch diameter divided by the lead.	$\cot A = \dfrac{3.1416\,D'}{L}$
Width of Thread Tool at Small End	Linear pitch	Multiply linear pitch by .31. Note: For 29° worm thread only.	$T = .31\,P$
Normal Thickness of Tooth at Pitch Line	Normal pitch	Divide normal pitch by 2.	$B = \dfrac{P'}{2}$
	Linear pitch Helix angle of worm	Multiply one half of the linear pitch by cosine of helix angle.	$B = \dfrac{P}{2}(\cos A)$
Whole Depth of Thread Tooth	Circular or linear pitch	Multiply circular pitch by .6866. Note: For helix angles less than 15°.	$W = .6866\,P$
	Normal pitch	Multiply normal pitch by .6866. Note: For helix angles greater than 15°.	$W = .6866\,P'$
Root Diameter of Worm	Outside diameter of worm Whole depth	Subtract twice the whole depth of tooth from the outside diameter.	$R = O' - 2W$
Face Width	Circular pitch	Multiply circular pitch by 2.38, then add .250. Note: Recommended for single and double threads by A.G.M.A.	$F = 2.38P + .250$
		Multiply circular pitch by 2.15, then add .200. Note: Recommended for triple and quadruple threads by A.G.M.A.	$F = 2.15P + .200$
Corner Radius	Circular pitch	Multiply circular pitch by .25.	$E = .25P$
Center Distance Between Worm and Gear	Pitch diameters of both gear and worm	Add pitch diameter of gear to pitch diameter of worm and divide by 2.	$C = \dfrac{D + D'}{2}$
Minimum Threaded Length of Worm	Pitch diameter of worm gear Addendum	Extract the square root of 8 times the pitch diameter multiplied by the addendum.	$X = \sqrt{8DS}$

The number of threads refers to the multiple of the worm and *not* to the number of threads per inch. For example, a single-threaded worm has *one* continuous thread, and a quadruple-threaded worm has *four* threads. Refer to Figure 21-2.

Normal pitch is the distance from a given point on one tooth on the worm or gear to the corresponding point on the next tooth, measured perpendicular to the helix angle or the side of the tooth.

Normal thickness of thread refers to the thickness of the tooth at the pitch line, measured perpendicular to the helix angle, or side of the tooth.

The *helix angle* of the worm is made by the helix of the thread at the pitch diameter with a plane perpendicular to the axis. The helix angle varies with the lead of the worm. The helix angle of the worm gear is equal to the lead angle of the worm.

The *linear pitch* of the worm is equal to the circular pitch of the worm gear.

The *throat diameter* on a worm gear is measured on the central plane which cuts through the lowest point of the concave rim, and is perpendicular to its axis.

Throat radius refers to the curvature at the throat and is equal to the radius of the worm at the working depth.

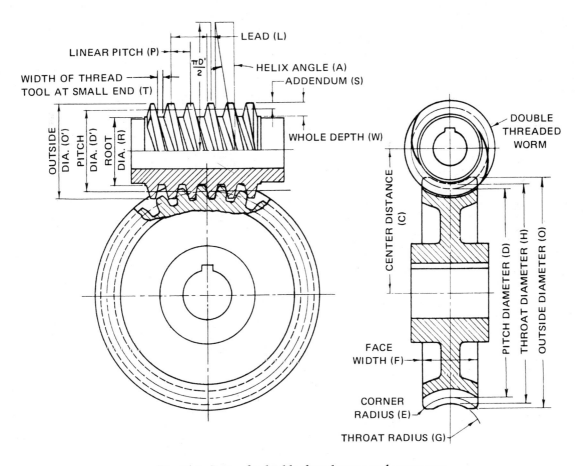

Fig. 21-2 **Parts of a double-thread worm and worm gear**

Unit 21 Worm Gearing

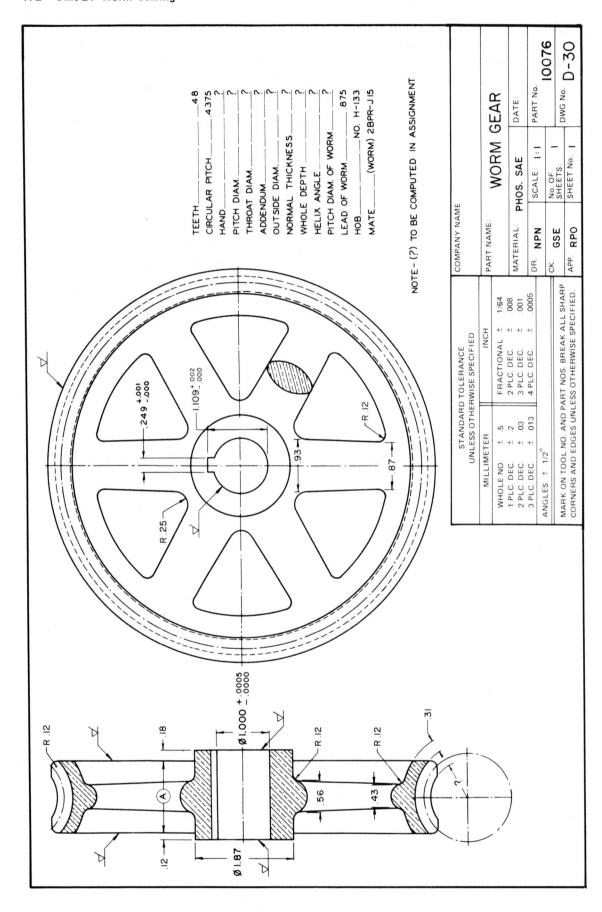

ASSIGNMENT D-30: WORM GEAR

1. Give the lead of the mating worm. _____
2. What is the linear pitch of the worm? _____
3. Give the pitch diameter of the worm. _____
4. Give the lead angle of the worm. _____
5. What is the difference between the lead angle of the worm and the helix angle of the Worm Gear? _____
6. State whether the Worm Gear teeth are cut for a right- or a left-hand worm. _____
7. What is the width of thread at the small end? _____
8. Determine the normal thickness of the worm thread at the pitch line. _____
9. Determine the whole depth of tooth. _____
10. What is the root diameter of the worm? _____
11. How many teeth are there in the Worm Gear? _____
12. What is the circular pitch of the Worm Gear? _____
13. Determine the addendum of the Worm Gear. _____
14. Determine the pitch diameter of the Worm Gear. _____
15. Determine the throat diameter. _____
16. Compute the outside diameter of the Worm Gear according to AGMA recommendations. _____
17. Determine the throat radius. _____
18. Determine the center distance between the worm and Worm Gear. _____
19. Compute the minimum threaded length of worm permissible. _____
20. Determine the width and depth of the keyway cut in the center hole of the Worm Gear. _____

Unit 21 Worm Gearing

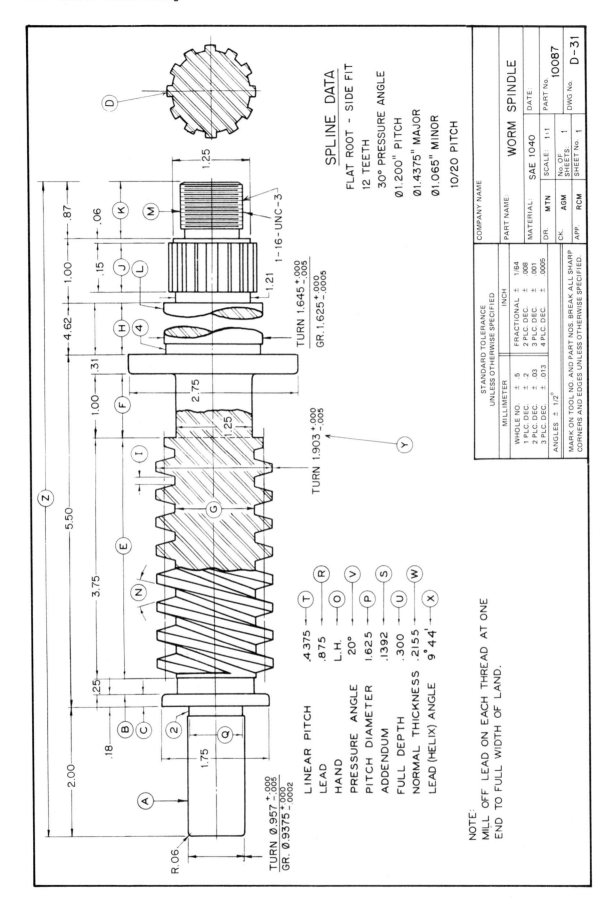

SPLINE DATA
FLAT ROOT - SIDE FIT
12 TEETH
30° PRESSURE ANGLE
Ø1.200" PITCH
Ø1.4375" MAJOR
Ø1.065" MINOR
10/20 PITCH

LINEAR PITCH	.4375
LEAD	.875
HAND	L.H.
PRESSURE ANGLE	20°
PITCH DIAMETER	1.625
ADDENDUM	.1392
FULL DEPTH	.300
NORMAL THICKNESS	.2155
LEAD (HELIX) ANGLE	9° 44'

NOTE:
MILL OFF LEAD ON EACH THREAD AT ONE END TO FULL WIDTH OF LAND.

COMPANY NAME
PART NAME: **WORM SPINDLE**
MATERIAL: SAE 1040
DR. MTN SCALE: 1:1 PART No. 10087
CK. AGM No. OF SHEETS: 1 DWG No. D-31
APP. RCM SHEET No. 1 DATE:

STANDARD TOLERANCE UNLESS OTHERWISE SPECIFIED

MILLIMETER		INCH	
WHOLE NO. ± .5		FRACTIONAL ± 1/64	
1 PLC. DEC. ± .2		2 PLC. DEC. ± .008	
2 PLC. DEC. ± .03		3 PLC. DEC. ± .001	
3 PLC. DEC. ± .013		4 PLC. DEC. ± .0005	

ANGLES ± 1/2°

MARK ON TOOL NO. AND PART NOS. BREAK ALL SHARP CORNERS AND EDGES UNLESS OTHERWISE SPECIFIED.

ASSIGNMENT D-31: WORM SPINDLE

1. What material is specified for this part?
2. How many thousandths are left at Ⓐ for grinding after the turning operation?
3. What is the size of the thread at Ⓜ?
4. What does the *16* stand for in the thread notation?
5. What do the *U*, *N*, and *C* stand for in the thread notation?
6. What does the *3* stand for in the thread notation?
7. The view at Ⓓ apparently indicates a sectional view. Where is this section taken?
8. How much oversize is Ⓛ turned?
9. Is dimension Ⓟ used in the machining of the part?
10. What is the number of threads per inch of the thread at Ⓔ?
11. What is angle Ⓝ?
12. What does the *LH* at Ⓞ stand for?
13. What are six working dimensions that must be known to cut the worm thread at Ⓔ?
14. To what length would the rough stock be cut for making this part?
15. What is diameter Ⓖ?
16. What given dimensions, other than Ⓟ and Ⓢ, are not used in cutting the worm thread at Ⓔ?
17. What is the ratio of linear pitch to the lead of the worm thread at Ⓔ?
18. What does the ratio of the linear pitch to the lead of the worm thread indicate?
19. What is the operation called to make the recesses at ② and ④?
20. Determine distance Ⓘ.

APPENDIX

Table A-1 Decimal Equivalent of Tool Sizes

Inch	Decimal	Wire	mm	Tap Sizes To be used with drills as indicated	Inch	Decimal	Wire	mm	Tap Sizes To be used with drills as indicated	Inch	Decimal	Wire	mm	Tap Sizes To be used with drills as indicated
	.0059	97	.15		1/32	.0312					.0827		2.1	
	.0063	96	.16			.0315		.8			.0846		2.15	
	.0067	95	.17			.0320	67				.0860	44		
	.0071	94	.18			.0330	66				.0866		2.2	
	.0075	93	.19			.0335		.85			.0886		2.25	
	.0079	92	.20			.0350	65				.0890	43		4-40
	.0083	91	.21			.0354		.9			.0906		2.3	
	.0087	90	.22			.0360	64				.0925		2.35	
	.0091	89	.23			.0370	63				.0935	42		4-48
	.0094		.24			.0374		.95		3/32	.0938			
	.0095	88				.0380	62				.0945		2.4	
	.0098		.25			.0390	61				.0960	41		
	.0100	87				.0394		1.			.0965		2.45	
	.0102		.26			.0400	60				.0980	40		
	.0105	86				.0410	59				.0984		2.5	
	.0106		.27			.0413		1.05			.0995	39		
	.0110	85	.28			.0420	58				.1015	38		5-40
	.0114		.29			.0430	57				.1024		2.6	
	.0115	84				.0433		1.1			.1040	37		5-44
	.0118		.30			.0453		1.15	0.80		.1063		2.7	
	.0120	83									.1065	36		6-32
	.0125	82				.0465	56				.1083		2.75	
	.0126		.32		3/64	.0469					.1094			
	.0130	81				.0472		1.2		7/64	.1100	35		
	.0134		.34			.0492		1.25			.1102		2.8	
	.0135	80				.0512		1.3			.1110	34		
	.0138		.35			.0520	55				.1130	33		6-40
	.0142		.36			.0531		1.35			.1142		2.9	
	.0145	79				.0550	54				.1160	32		
	.0150		.38			.0551		1.4			.1181		3.	
1/64	.0156					.0571		1.45			.1200	31		
	.0157		.4			.0591		1.5	1-64		.1220		3.1	
	.0160	78				.0595	53		1-72		.1250			
	.0165		.42			.0610		1.55		1/8	.1260		3.2	
	.0173		.44		1/16	.0625					.1280		3.25	
	.0177		.45			.0630		1.6			.1285	30		
	.0180	77				.0635	52				.1299		3.3	
	.0181		.46			.0650		1.65			.1339		3.4	
	.0189		.48			.0669		1.7			.1360	29		8-32
	.0197		.5			.0670	51				.1378		3.5	8-36
	.0200	76				.0689		1.75	2-56		.1405	28		
	.0210	75				.0700	50		2-64		.1406			
	.0217		.55			.0709		1.8		9/64	.1417		3.6	
	.0225	74				.0728		1.85			.1440	27		
	.0236		.6			.0730	49				.1457		3.7	
	.0240	73				.0748		1.9			.1470	26		
	.0250	72				.0760	48				.1476		3.75	
	.0256		.65			.0768		1.95			.1495	25		10-24
	.0260	71			5/64	.0781					.1496		3.8	
	.0276		.7			.0785	47		3-48		.1520	24		
	.0280	70				.0787		2.			.1535		3.9	
	.0292	69				.0807		2.05			.1540	23		
	.0295		.75			.0810	46			5/32	.1562			
	.0310	68				.0820	45		3-56					

Table A-1 Decimal Equivalent of Tool Sizes (continued)

Inch	Decimal	Wire and Letter	mm	Tap Sizes To be used with drills as indicated	Inch	Decimal	Letter	mm	Tap Sizes To be used with drills as indicated	Inch	Decimal	Letter	mm	Tap Sizes To be used with drills as indicated
	.1570	22				.2362		6.			.3386		8.6	
	.1575		4.			.2380	B				.3390	R		
	.1590	21		10-32		.2402		6.1			.3425		8.7	
	.1610	20				.2420	C			11/32	.3438			
	.1614		4.1			.2441		6.2			.3445		8.75	
	.1654		4.2			.2460	D				.3465		8.8	
	.1660	19				.2461		6.25			.3480	S		
	.1673		4.25			.2480		6.3			.3504		8.9	
	.1693		4.3		1/4	.2500	E				.3543		9.	
	.1695	18				.2520		6.4			.3580	T		
11/64	.1719					.2559		6.5			.3583		9.1	
	.1730	17				.2570	F		5/16-18	23/64	.3594			
	.1732		4.4			.2598		6.6			.3622		9.2	
	.1770	16		12-24		.2610	G				.3642		9.25	
	.1772		4.5			.2638		6.7			.3661		9.3	
	.1800	15			17/64	.2656					.3680	U		7/16-14
	.1811		4.6			.2657		6.75			.3701		9.4	
	.1820	14		12-28		.2660	H				.3740		9.5	
	.1850	13				.2677		6.8		3/8	.3750			
	.1850		4.7			.2717		6.9			.3770	V		
	.1870		4.75			.2720	I		5/16-24		.3780		9.6	
3/16	.1875					.2756		7.			.3819		9.7	
	.1890		4.8			.2770	J				.3839		9.75	
	.1890	12				.2795		7.1			.3858		9.8	
	.1910	11				.2810	K				.3860	W		
	.1929		4.9		9/32	.2812					.3898		9.9	
	.1935	10				.2835		7.2		25/64	.3906			7/16-20
	.1960	9				.2854		7.25			.3937		10.	
	.1969		5.			.2874		7.3			.3970	X		
	.1990	8				.2900	L				.4040	Y		
	.2008		5.1			.2913		7.4		13/32	.4062			
	.2010	7		1/4-20		.2950	M				.4130	Z		
13/64	.2031					.2953		7.5			.4134		10.5	
	.2040	6			19/64	.2969					.4219			1/2-13
	.2047		5.2			.2992		7.6		27/64	.4331		11.	
	.2055	5				.3020	N			7/16	.4375			
	.2067		5.25			.3031		7.7						
	.2087		5.3			.3051		7.75			.4528		11.5	
	.2090	4				.3071		7.8		29/64	.4531			1/2-20
	.2126		5.4			.3110		7.9						
	.2130	3		1/4-28	5/16	.3125			3/8-16	15/32	.4688			
	.2165		5.5			.3150		8.			.4724		12.	
7/32	.2188					.3160	O							9/16-12
	.2205		5.6			.3189		8.1		31/64	.4844			
	.2210	2				.3228		8.2			.4921		12.5	
	.2244		5.7			.3230	P			1/2	.5000			
	.2264		5.75			.3248		8.25			.5118		13.	
	.2280	1				.3268		8.3		33/64	.5156			9/16-18
	.2283		5.8		21/64	.3281				17/32	.5312			5/8-11
	.2323		5.9			.3307		8.4			.5315		13.5	
	.2340	A				.3320	Q		3/8-24	35/64	.5469			
15/64	.2344					.3346		8.5			.5512		14.	

Appendix 177

Table A-1 Decimal Equivalent of Tool Sizes (continued)

Inch	Decimal	mm	Tap Sizes To be used with drills as indicated	Inch	Decimal	mm	Tap Sizes To be used with drills as indicated	Inch	Decimal	mm	Tap Sizes To be used with drills as indicated
9/16	.5625				.9843	25.		1 25/64	1.3906		
	.5709	14.5		63/64	.9844		1 1/8-7		1.3976	35.5	
37/64	.5781		5/8-18	1	1.0000			1 13/32	1.4062		
	.5906	15.			1.0039	25.5			1.4173	36.	
19/32	.5938			1 1/64	1.0156			1 27/64	1.4219		1 1/2-12
39/64	.6094				1.0236	26.			1.4370	36.5	
	.6102	15.5		1 1/32	1.0312			1 7/16	1.4375		
5/8	.6250				1.0433	26.5		1 29/64	1.4531		
	.6299	16.		1 3/64	1.0469		1 1/8-12		1.4567	37.	
41/64	.6406			1 1/16	1.0625			1 15/32	1.4688		
	.6496	16.5			1.0630	27.			1.4764	37.5	
21/32	.6562		3/4-10	1 5/64	1.0781			1 31/64	1.4844		
	.6693	17.			1.0827	27.5			1.4961	38.	
43/64	.6719			1 3/32	1.0938			1 1/2	1.5000		
11/16	.6875		3/4-16		1.1024	28.		1 33/64	1.5156		
	.6890	17.5		1 7/64	1.1094		1 1/4-7		1.5157	38.5	
45/64	.7031				1.1220	28.5		1 17/32	1.5312		
	.7087	18.		1 1/8	1.1250				1.5354	39.	
23/32	.7188			1 9/64	1.1406			1 35/64	1.5469		
	.7283	18.5			1.1417	29.			1.5551	39.5	
47/64	.7344			1 5/32	1.1562			1 9/16	1.5625		
	.7480	19.			1.1614	29.5			1.5748	40.	
3/4	.7500			1 11/64	1.1719		1 1/4-12	1 37/64	1.5781		
49/64	.7656		7/8-9		1.1811	30.		1 19/32	1.5938		
	.7677	19.5		1 3/16	1.1875				1.5945	40.5	
25/32	.7812				1.2008	30.5		1 39/64	1.6094		
	.7874	20.		1 13/64	1.2031				1.6142	41.	
51/64	.7969			1 7/32	1.2188		1 3/8-6	1 5/8	1.6250		
	.8071	20.5			1.2205	31.			1.6339	41.5	
13/16	.8125		7/8-14	1 15/64	1.2344			1 41/64	1.6406		
	.8268	21.			1.2402	31.5			1.6535	42.	
53/64	.8281			1 1/4	1.2500			1 21/32	1.6562		
27/32	.8438				1.2598	32.		1 43/64	1.6719		
	.8465	21.5		1 17/64	1.2656				1.6732	42.5	
55/64	.8594				1.2795	32.5		1 11/16	1.6875		
	.8661	22.		1 9/32	1.2812				1.6929	43.	
7/8	.8750		1-8	1 19/64	1.2969		1 3/8-12				
	.8858	22.5			1.2992	33.		1 45/64	1.7031		
57/64	.8906			1 5/16	1.3125				1.7126	43.5	
	.9055	23.						1 23/32	1.7188		
29/32	.9062		1-12		1.3189	33.5			1.7323	44.	
59/64	.9219		1-14	1 21/64	1.3281						
	.9252	23.5			1.3386	34.		1 47/64	1.7344		
15/16	.9375			1 11/32	1.3438		1 1/2-6	1 3/4	1.7500		
	.9449	24.			1.3583	34.5			1.7520	44.5	
61/64	.9531			1 23/64	1.3594			1 49/64	1.7656		
	.9646	24.5		1 3/8	1.3750				1.7717	45.	
31/32	.9688				1.3780	35.		1 25/32	1.7812		

Table A-1 Decimal Equivalent of Tool Sizes (continued)

Inch	Decimal	mm	Tap Sizes To be used with drills as indicated	Inch	Decimal	mm	Tap Sizes To be used with drills as indicated	Inch	Decimal	mm	Tap Sizes To be used with drills as indicated
	1.7913	45.5			2.2047	56.			2.6181	66.5	
1 51/64	1.7969			2 7/32	2.2188			2 5/8	2.6250		
	1.8110	46.			2.2244	56.5			2.6378	67.	
1 13/16	1.8125			2 15/64	2.2344			2 41/64	2.6406		
1 53/64	1.8281				2.2441	57.		2 21/32	2.6562		
	1.8307	46.5		2 1/4	2.2500				2.6575	67.5	
1 27/32	1.8438				2.2638	57.5		2 43/64	2.6719		
	1.8504	47.		2 17/64	2.2656				2.6772	68.	
1 55/64	1.8594			2 9/32	2.2812			2 11/16	2.6875		
	1.8701				2.2835	58.			2.6969		
		47.5								68.5	
1 7/8	1.8750			2 19/64	2.2969			2 45/64	2.7031		
	1.8898	48.			2.3032	58.5			2.7165	69.	
1 57/64	1.8906			2 5/16	2.3125			2 23/32	2.7188		
1 29/32	1.9062				2.3228	59.		2 47/64	2.7344		
	1.9094	48.5		2 21/64	2.3281				2.7362	69.5	
1 59/64	1.9219				2.3425	59.5		2 3/4	2.7500		
	1.9291	49.		2 11/32	2.3438				2.7559	70.	
1 15/16	1.9375			2 23/64	2.3594			2 49/64	2.7656		
	1.9488	49.5			2.3622	60.			2.7756	70.5	
1 61/64	1.9531			2 3/8	2.3750			2 25/32	2.7812		
	1.9685	50.			2.3819	60.5			2.7953	71.	
1 31/32	1.9688			2 25/64	2.3906			2 51/64	2.7969		
1 63/64	1.9844				2.4016	61.		2 13/16	2.8125		
	1.9882	50.5		2 13/32	2.4062				2.8150	71.5	
2	2.0000				2.4213	61.5					
	2.0079	51.		2 27/64	2.4219			2 53/64	2.8281		
2 1/64	2.0156								2.8346	72.	
	2.0276	51.5		2 7/16	2.4375			2 27/32	2.8438		
2 1/32	2.0312				2.4409	62.			2.8543	72.5	
2 3/64	2.0469			2 29/64	2.4531						
	2.0472	52.			2.4606	62.5		2 55/64	2.8594		
2 1/16	2.0625								2.8740	73.	
	2.0669	52.5		2 15/32	2.4688			2 7/8	2.8750		
2 5/64	2.0781				2.4803	63.		2 57/64	2.8906		
	2.0866	53.		2 31/64	2.4844						
2 3/32	2.0938			2 1/2	2.5000	63.5			2.8937	73.5	
								2 29/32	2.9062		
	2.0163	53.5		2 33/64	2.5156				2.9134	74.	
2 7/64	2.1094				2.5197	64.		2 59/64	2.9219		
2 1/8	2.1250			2 17/32	2.5312						
	2.1260	54.			2.5394	64.5			2.9331	74.5	
								2 15/16	2.9375		
2 9/64	2.1406			2 35/64	2.5469				2.9528	75.	
	2.1457	54.5			2.5591	65.		2 61/64	2.9531		
2 5/32	2.1562			2 9/16	2.5625						
	2.1654	55.		2 37/64	2.5781			2 31/32	2.9688		
									2.9724	75.5	
2 11/64	2.1719				2.5787	65.5		2 63/64	2.9844		
	2.1850	55.5		2 19/32	2.5938				2.9921	76.	
2 3/16	2.1875				2.5984	66.					
2 13/64	2.2031			2 39/64	2.6094			3	3.0000		

Appendix 179

Table A-1 Metric

Metric Tap Size	RECOMMENDED METRIC DRILL				CLOSEST RECOMMENDED INCH DRILL			
	Drill Size mm	Inch Equiv.	Probable Hole Size (Inches)	Probable Percent of Thread	Drill Size	Inch Equiv.	Probable Hole Size (Inches)	Probable Percent of Thread
M1.6 X 0.35	1.25	0.0492	0.0507	69	—	—	—	—
M1.8 X 0.35	1.45	0.0571	0.0586	69	—	—	—	—
M2 X 0.4	1.60	0.0630	0.0647	69	#52	0.0635	0.0652	66
M2.2 X 0.45	1.75	0.0689	0.0706	70	—	—	—	—
M2.5 X 0.45	2.05	0.0807	0.0826	69	#46	0.0810	0.0829	67
M3 X 0.5	2.50	0.0984	0.1007	68	#40	0.0980	0.1003	70
M3.5 X 0.6	2.90	0.1142	0.1168	68	#33	0.1130	0.1156	72
M4 X 0.7	3.30	0.1299	0.1328	69	#30	0.1285	0.1314	73
M4.5 X 0.75	3.70	0.1457	0.1489	74	#26	0.1470	0.1502	70
M5 X 0.8	4.20	0.1654	0.1686	69	#19	0.1660	0.1692	68
M6 X 1	5.00	0.1968	0.2006	70	#9	0.1960	0.1998	71
M7 X 1	6.00	0.2362	0.2400	70	15/64	0.2344	0.2382	73
M8 X 1.25	6.70	0.2638	0.2679	74	17/64	0.2656	0.2697	71
M8 X 1	7.00	0.2756	0.2797	69	J	0.2770	0.2811	66
M10 X 1.5	8.50	0.3346	0.3390	71	Q	0.3320	0.3364	75
M10 X 1.25	8.70	0.3425	0.3471	73	11/32	0.3438	0.3483	71
M12 X 1.75	10.20	0.4016	0.4063	74	Y	0.4040	0.4087	71
M12 X 1.25	10.80	0.4252	0.4299	67	27/64	0.4219	0.4266	72
M14 X 2	12.00	0.4724	0.4772	72	15/32	0.4688	0.4736	76
M14 X 1.5	12.50	0.4921	0.4969	71	—	—	—	—
M16 X 2	14.00	0.5512	0.5561	72	35/64	0.5469	0.5518	76
M16 X 1.5	14.50	0.5709	0.5758	71	—	—	—	—
M18 X 2.5	15.50	0.6102	0.6152	73	39/64	0.6094	0.6144	74
M18 X 1.5	16.50	0.6496	0.6546	70	—	—	—	—
M20 X 2.5	17.50	0.6890	0.6942	73	11/16	0.6875	0.6925	74
M20 X 1.5	18.50	0.7283	0.7335	70	—	—	—	—
M22 X 2.5	19.50	0.7677	0.7729	73	49/64	0.7656	0.7708	75
M22 X 1.5	20.50	0.8071	0.8123	70	—	—	—	—
M24 X 3	21.00	0.8268	0.8327	73	53/64	0.8281	0.8340	72
M24 X 2	22.00	0.8661	0.8720	71	—	—	—	—
M27 X 3	24.00	0.9449	0.9511	73	15/16	0.9375	0.9435	78
M27 X 2	25.00	0.9843	0.9913	70	63/64	0.9844	0.9914	70
M30 X 3.5	26.50	1.0433						
M30 X 2	28.00	1.1024						
M33 X 3.5	29.50	1.1614						
M33 X 2	31.00	1.2205	REAMING RECOMMENDED TO THE DRILL SIZE SHOWN						
M36 X 4	32.00	1.2598						
M36 X 3	33.00	1.2992						
M39 X 4	35.00	1.3780						
M39 X 3	36.00	1.4173						

FORMULA FOR METRIC TAP DRILL SIZE

$$\text{Basic Major Dia. (mm)} - \frac{\% \text{ Thread} \times \text{Pitch (mm)}}{76.980} = \text{DRILLED HOLE SIZE (mm)}$$

FORMULA FOR PERCENT OF THREAD

$$\frac{76.980}{\text{Pitch (mm)}} \times \left[\text{Basic Major Dia. (mm)} - \text{Drilled Hole Size (mm)} \right] = \text{Percent of Thread}$$

TABLE A-2: Speeds for High Speed Steel Drills

DRILL SIZE		STEEL CASTING		TOOL STEEL		CAST IRON		MACHINE STEEL		BRASS AND ALUMINUM	
		CUTTING SPEEDS IN METRES PER MINUTE OR FEET PER MINUTE									
mm	in.	12 m/min	40 ft./min	18 m/min	60 ft./min	24 m/min	80 ft./min	30 m/min	100 ft./min	60 m/min	200 ft./min
		Revolution Per Minute									
2	1/16	1910	2445	2865	3665	3820	4890	4775	6110	9550	12225
3	1/8	1275	1220	1910	1835	2545	2445	3185	3055	6365	6110
4	3/16	955	815	1430	1220	1910	1630	2385	2035	4775	4075
5	1/4	765	610	1145	915	1530	1220	1910	1530	3820	3055
6	5/16	635	490	955	735	1275	980	1590	1220	3180	2445
7	3/8	545	405	820	610	1090	815	1365	1020	2730	2035
8	7/16	475	350	715	525	955	700	1195	875	2390	1745
9	1/2	425	305	635	460	850	610	1060	765	2120	1530
10	5/8	350	245	520	365	695	490	870	610	1735	1220
15	3/4	255	205	380	305	510	405	635	510	1275	1020
20	7/8	190	175	285	260	380	350	475	435	955	875
25	1"	150	155	230	230	305	305	380	380	765	765

ILLUSTRATION CONTRIBUTIONS

From *Blueprint Reading for Machinists – Advanced.* © 1972 by Delmar Publishers Inc. – Figures 5-6, 5-7, 21-1, 21-2. Tables 21-1, 21-2. Drawing Assignment D-30.

From *Blueprint Reading for Machinists – Intermediate.* © 1971 by Delmar Publishers Inc. – Figures 1-10, 1-11, 2-2, 2-5, 2-6, 4-1, 4-2, 4-3, 4-4, 4-5, 4-6, 4-7, 5-1, 5-2, 5-3, 5-4, 5-5, 5-8, 6-2, 9-7, 9-8, 9-9, 9-10, 9-13, 13-1, 13-2, 13-3, 13-4, 13-5, 14-3, 14-7, 14-9, 14-10, 14-11, 14-12, 16-2, 16-3, 17-7, 18-1, 18-2, 18-3, 18-4, 18-7, 19-1, 19-2, 19-4, 20-1, 20-2, 20-3, 20-4, 20-5, 20-6. Tables 11-1, 12-1, 12-2, 12-3. Drawing Assignments D-1, D-3, D-4, D-5, D-6, D-7, D-8, D-9, D-10, D-11, D-14, D-15, D-16, D-18, D-19, D-20, D-22, D-23, D-24, D-25, D-26, D-27, D-28, D-29, D-31.

From C. Jensen & R. Hines, *Interpreting Engineering Drawings.* © 1970 by Delmar Publishers Inc. – Figures 1-13, 1-14, 1-15, 6-4, 6-5. Drawing Assignments D-12, D-13.

From C. Jensen & R. Hines, *Interpreting Engineering Drawings, Metric Edition.* © 1979 by Delmar Publishers Inc. – Figures 3-4, 3-6.

From C. T. Olivo, *Advanced Machine Technology.* © 1982 by Breton Publishers. – Table 3-1.

From David L. Taylor, *Drill Press Work.* © 1980 by Delmar Publishers Inc. – Appendix Tables A-1, A-2.

From David L. Taylor, *Elementary Blueprint Reading for Machinists.* © 1981 by Delmar Publishers Inc. – Figures 1-1, 1-2, 1-3, 1-4, 1-5, 1-6, 1-7, 1-8, 1-9, 2-1, 9-11, 9-12, 9-14.

From David L. Taylor, *Machine Trades Blueprint Reading.* © 1984 by Delmar Publishers Inc. – Figures 1-12, 3-1, 3-2, 3-3, 3-5, 6-1, 6-3A, 6-3B, 7-1, 7-2, 7-3, 7-4, 7-5, 7-6, 7-7, 7-9, 7-10, 8-1, 8-2, 8-3, 8-4, 8-5, 8-6, 8-7, 9-1, 9-2, 9-3, 9-4, 9-5, 9-6, 10-1, 10-2, 10-3, 10-4, 10-5, 10-6, 10-7, 11-1, 14-1, 14-2, 14-4, 14-5, 14-6, 14-8, 17-1, 17-2, 17-6. Drawing Assignments D-17, D-21.

American Institute of Steel Construction – Table 16-1.

American National Standards Institute – Figures 2-3, 18-5. Tables 10-1, 10-2, 18-1, 18-2, 18-3, 18-4, 18-5, 18-6.

American Society of Mechanical Engineers – Figures 2-3, 18-5. Tables 10-1, 10-2, 18-1, 18-2, 18-3, 18-4, 18-5, 18-6.

American Welding Society – Figures 17-3, 17-4, 17-5.

Blackstone Corporation – Drawing Assignment D-17.

Cummins Engine Company – Drawing Assignment D-2.

Eldon Peterson – Worksheets for Geometric Positioning and Tolerancing (Unit 8).

INDEX

Aligned projection, 32, 33
Auxiliary pump base, 124-25

Base plate, 12-13
Bolts, types, 89
Bottom views, 43
Break lines, 41
Broken-out sections, 18, 19
Built-up sections, structural steel, 133

CAD for drawings, 8
Cap screws, 88, 89
Case cover, 56-57
Castings, cored, 122-23
    advantages, 122
    appearance, 122
    and holes, 122
    reasons for, 122-23
    section, 123
Clevis pins, 151-52
    diagram, 152
Coil frame, 38-39
Conversion coating, 126-27
    flow, 126
    iron phosphate, 126
    manganese phosphate, 127
    nature, 126
    zinc phosphate, 126
Coring, 121-22
    discussion, 121
    process, 122
    ready for baking, 121
    removal of, 122
Corner bracket, 128-29
Cotter pins, 155
Cross head, 85-86
Cutting planes:
    line, 14
    locating, 14
    reasons for, 14

Datum axis, 60, 61
Datum cylinder, 60, 61
Datum, defined, 59
Datum dimensioning, 51-53
    centerline, 53
    curved lines, 53

    data points, 51
    diagram, 52
    discussion, 51
    machined edges, 53
    multiple data points, 52
Datum identification symbol, 61
    and extension lines, 61
Datum planes, 59-60
    discussion, 59
    primary, 59
    secondary, 60
    tertiary, 60
Distorted views:
    example, 42, 43
    lugs, 43
    reasons for, 42
    vs. true, 42, 43
Dovetails:
    corners, 112
    dimensions, 111
    discussion, 111-12
    female, 112
    formulas, 112
    male, 113
    measuring, 112
    parts, 111
Dowel pins, 148-50
    aligning parts with, 148
    discussion, 148
    hardened ground machine, 149
        chart, 149
        diagram, 149
    hardened ground production, 150
        chart, 150
        diagram, 150
Drill slide, 116-17
Drive housing, 102-03

Electroplating, 127

Fasteners, typical assemblies, 93
Feature control symbols, 61-63
    chart, 62
    frame, 63
    leader line, 63
First angle projection, 7
Fits of screw threads, classes, 81

Flamespray coating, 127
Flange, 96-97
Flat back patterns, 119, 121
    discussion, 119
    example, 121
    mold for, making, 121
Fluid pressure valve, 162-63
Form tolerances:
    angularity, 71
    cylindricity, 69
    diagrams, 69-70
    flatness, 68
    parallelism, 71
    perpendicularity, 71
    profile of a line, 69
    profile of a surface, 71
    roundness, 69
    runout, 71
    straightness, 69
Four-wheel trolley, 134-35
Full sections, 17

Grooved pins, 150-51
    dimensions, 152
    discussion, 150
    types, 151

Half sections, 18
Holes, 8, 9
Hood, 156-57

Index pedestal, 46-47
Interlock base, 54-55
ISO projection symbols, 7-8

Lay, of surface, symbols for, 26-27
Location tolerances:
    concentricity, 68
    diagram, 68
    position, 68

Machined parts, reasons for finishing, 126
Machine screws, 88
Magazine bar feed stock pusher guide, 144-45
Maximum material condition:
    diagrams, 67
    discussion, 67

184 Index

Microinch ratings of surfaces:
  by manufacturing processes, 28, 29
  typical applications, 28
Micrometer surface ranges, by process, 29
Multiple threads, screw:
  diagrams, 81
  discussion, 81
  reasons for, 81

Nuts, types, 89

Offset cutting planes, 17
Organic coatings, 127

Partial sections, 17
Partial views, 41
Phantom lines and views, 44-45
  alternate positions of moving parts, 45
  discussion, 44
  machining lugs, 44
  repetitive features, 45
  use, 44
Pipe threads:
  American National Standard, 99, 101
    table, 101
  modified, special-purpose, 99
  representing of, 100
  specification, 100
  straight, 99-100
  symbols, 99
  tapered, 99
Point-to-point dimensions, 51
Positioning arm, 64-65
Profilometers, 27
Projection lines, 5
  right view, use for, 5
Projection, orthographic:
  discussion, 3
  front, 1, 2
  side, 1, 3
  top, 1, 2
  transparent-sided box, 2
  working drawing, 1

Raise block, 20-21
Rear tool post, 108-09
Regardless of feature size, 67-68
Removed sections, 18, 19
Repetitive features, specification of, 8
  and holes, 8
Revolved sections, 18
Rotated projection, 32, 33
  and section views, 33

Sand molding:
  casting from, 120
  closed flask for, 118
  core, 120
  gate, 119
  opened flask, 118
  shrinkage, 119
  sprue, 119
  wood pattern for, 119
Screw threads, forms of:
  Acme, 78
  American National, 78
  Buttress, 78
  diagram, 77
  Rolled, 78
  Sharp V, 77
  Unified, 77, 78
  Whitworth, 78
  Worm, 78
Screw thread terminology, 78-79
  helix thread, 78
  notation, 79
  nut, 79
  specification, 79
Screws, representation of:
  pictorial, 82
  schematic:
    discussion, 82
    external, 82
    internal, 83
    simplified, 83
  simplified, 83
  tapped, 83-84
    depth of threading, 84
    diagrams, 83, 84
    nature, 83, 84
    representation, 84
    simplified, 84
Section lines, 17
Section lining, 15
Sections, types of, 17
Separator bracket, 10-11
Setscrews, 90
Shaft intermediate support, 30-31
Shuttle, 114-15
Side views:
  full section, 16
  half section, 16
  not sectioned, 16
Slide valve, 22-23
Spark adjuster, 36-37
Spider, 94-95
Spindle bearing, 87
Split bearings:
  diagram, 165
  mismatch problem, 165
  reasons for, 165
  shimming, 165

Split bushings, 165
Spring pins:
  coiled type:
    diagram, 154
    dimensions, 154
    discussion, 151
  slotted type:
    diagram, 153
    dimensions, 153
Springs:
  flat, 160
  helical, 159, 160
  information on drawings, 161
  pitch, 161
  simplified helical, 160
  specification, 161
  types, 159
Steels, identification of:
  AISI system, 104, 106
  boron, 106
  carbon, free-cutting, SAE, 105
  carbon, plain, SAE, 105
  example, 104
  lead, 106
  prefixes, 106
  SAE system, 104, 105
  suffixes, 106, 107
Straight pins, 148, 150
Structural steel shapes:
  AISI/AISC designations, 131-32
  identification, 133
  shapes, chart, 131
Stud bolts, 89
Support assembly valves, 142-43
Surface texture, measuring of, 27
Surface texture:
  definition, 24
  diagram, 24
  positioning, 25
  symbols, 25
    lay, 26
  terminology, 24-25
Swivels, 164
Symmetrical object, half view, 42

Taper pins, 146-48
  applications, 146
  diagram, 146
  dimension chart, 147
  dimension diagram, 147
  discussion, 146, 148
  drilling of holes for, 146, 148
  misalignment problem, 148
  standard taper in, 146
Third angle projection, 5-7
  nature, 5, 7
  relative position of views in, 6
  viewing plane, 6

Threaded fastener size:
  determining, 90
  specifications, 93
Thread specifications, symbols for, 82
Tolerance symbology, 71-73
Tolerances, geometric, 58
Trip box, 74-75

Unified National Thread series, screws:
  discussion, 80
  specifications, 80
Universal joint, 164
Universal trolley, 166-67
Untrue projection:
  diagrams, 34
  discussion, 33
  pulley, 35
  of sections, 34
  vs. true, 34
  webs and spokes, 34

Views, arrangement of, 4
Visualizing, 1

Washers, 90-92
  nature, 90
  regular helical, dimensions, 92
  Type A, preferred sizes, 91
Welding:
  chart, 138
  elements, 140
  and fabrication, 137
  fillet weld, dimensioning of, 141
  joints, 137
  location of symbols, 141
  supplementary symbols, 140
  symbols, 139
  terminology, 139, 141
  types, 137
Worm gearing:
  chamfered rim, 168, 169
  double thread, parts, 171
  gear, blueprint, 172-73
  helix angle, 171
  lead, 169
  normal pitch, 171
  normal thickness of thread, 171
  number of threads, 169, 171
  pitch, 169
  plane in, effects, 169
  representing, 168-69
  rims, 169
  rounded rim, 168, 169
  rules and formulas, gearing, 170
  skills required, 168
  spindle, blueprint, 174-75
  symbols and terminology table, 169
  throat diameter, 171
  throat radius, 171
  use, 168
  worm gear, nature, 168
  worm, nature, 168

Yoke, 48-49

4/99(7C1198K)